Point-of-use/ Point-of-entry

for Drinking Water Treatment

Benjamin W. Lykins, Jr.
Robert M. Clark
James A. Goodrich

LEWIS PUBLISHERS
Boca Raton   Ann Arbor   London

Library of Congress Cataloging-in-Publication Data

Lykins, Ben W.
        Point-of-use/point-of-entry for drinking water treatment/
    Benjamin W. Lykins, Jr., Robert M. Clark, and James A. Goodrich
        p.   cm.
      1. Drinking water--Purification--Equipment and supplies.
    2. Household appliances. I. Clark, Robert Maurice.   II.
    Goodrich, James A.   III. Title.
    TD433.L95 1991
    696'.12--dc20   91-26816
    ISBN 0-87371-354-0

LEWIS PUBLISHERS, INC.
121 South Main Street, Chelsea, Michigan 48118

PRINTED IN THE UNITED STATES OF AMERICA 1 2 3 4 5 6 7 8 9 0

This book is dedicated to my father.

Benjamin W. Lykins
1916 - 1991

"To everything there is a season ....
A time to be born, and a time to die ....
A time to laugh, and a time to mourn .... ."

Ecclesiastes 3:1-4

The public is more knowledgeable and concerned about contamination in their drinking water than at any other time in history. However, as the old saying goes, "a little knowledge can be dangerous". Sometimes this is true relative to point-of-use and point-of-entry treatment. When spills occur in water bodies used as sources of drinking water, the opportunity is available to prey on the public. Scare tactics and half-truths have been used to sell point-of-use and point-of-entry devices during these events. The public has been led to believe, at times, that the community water treatment is inadequate to produce safe drinking water. In most cases, this is not true. Point-of-use and point-of-entry devices used in homes on community drinking water systems are generally for aesthetic reasons. Fortunately, the point-of-use/point-of-entry industry is beginning to police itself and trying to control unscrupulous sales practices.

There are, however, several instances where point-of-use/point-of-entry devices are the most appropriate treatment for removing contaminants from drinking water. In these cases, the homeowner is encouraged to purchase point-of-use/point-of-entry devices. However, there is always the concern about proper operation and maintenance. The authors of this book neither encourage nor condemn the use of point-of-use/point-of-entry treatment. The authors have, however, tried to present information and data that will help in the decision process if one is considering point-of-use/point-of-entry treatment. This book presents provisions of the Safe Drinking Water Act which should be used as the basis for all drinking water quality, regardless of what treatment system or device is used. Types of point-of-use and point-of-entry devices are identified; application of these devices are shown; installation, operation, and maintenance suggestions are discussed; case studies are presented; cost of the devices are shown; State and Federal regulations are presented; and steps the consumer can take to protect themselves are identified.

Benjamin W. Lykins, Jr. has a BS in chemistry from Marshall University in Huntington, West Virginia (1963) and an MS in environmental engineering from the University of Cincinnati (1974).

Mr. Lykins is currently employed at the U.S. Environmental Protection Agency as Chief of the Systems and Field Evaluation Branch, Drinking Water Research Division, Risk Reduction Engineering Laboratory, Cincinnati. He is responsible for providing an integrated analysis of the cost and performance of technology associated with supplying safe drinking water to the public. This work involves conducting both intramural and extramural research. This research emphasizes both performance and costs associated with construction, operation and maintenance of modular and full scale water treatment system. Other research areas consist of predictive modeling, evaluating the factors affecting the deterioration of drinking water quality in water delivery systems, and small systems including point-of-use/point-of-entry.

Mr. Lykins has over 80 publications and presentations. He has served as a member of the American Water Works Association Academic Achievement Awards Committee and the Research Division Organic Contaminants Committee. He is also a member of several project advisory committees for the American Water Works Association Research Foundation. He is a member of the American Society of Civil Engineers' Environmental Engineering Division Point-of-Entry/Point-of-Use Committee and the Steering Committee for the Ontario Ministry of the Environment trace organics contaminant removal project.

Robert M. Clark has a BS in civil engineering from Oregon State University (1960) and in mathematics from Portland State University (1961). He also has an MS in mathematics from Xavier University in Cincinnati (1964) and in civil engineering from Cornell University (1968). He received his PhD from the University of Cincinnati in 1976.

Dr. Clark, a commissioned officer in the U.S. Public Health Service since 1961, is assigned to the U.S. Environmental Protection Agency as Director of the Drinking Water Research Division, Risk Reduction Engineering Laboratory, Cincinnati. He is responsible for directing and conducting research that meets the technology requirements for maintaining Maximum Contaminant Levels under the Safe Drinking Water Act. This work included formulating and guiding a broad-based program conducting research into frontier areas critically important to the provision of safe drinking water in the United States. Results of this research have had and will have a major impact on U.S. policies and procedures in water supply and on international drinking water practices.

Dr. Clark has received many professional awards, including selection by the American Society of Civil Engineers in 1980 as recipient of the Walter L. Huber Civil Engineering Prize. He received both the Commendation and Meritorious Service Awards from the U.S. Public Health Service (in 1977 and 1983, respectively). He has served as the chairman of the Executive Committee of the Environmental Engineering Division of the American Society of Civil Engineers and is currently vice-chairman of the Research Division Trustees for the American Water Works Association.

James A. Goodrich has a BA in Geography from the University of Cincinnati (1977), a MS in Geography from Florida State University (1978), and a Ph.D in Geography from the University of Cincinnati (1983).

Dr. Goodrich is currently employed at the U.S. Environmental Protection Agency as an Environmental Scientist for the Systems and Field Evaluation Branch, Drinking Water Research Division, Risk Reduction Engineering Laboratory, Cincinnati. Dr. Goodrich is responsible for managing research projects on the treatment of contaminated drinking water from hazardous waste sites, providing technical and institutional solutions to help small community water supplies be in compliance with the Safe Drinking Water Act and its amendments, and the impact drinking water distribution network operation and design has on water quality reaching the consumer. Additional research is underway evaluating the impact alternative disinfectants have on the water quality in distribution systems.

Dr. Goodrich has over 60 publications and presentations. He is a member of the American Water Works Association "Small Systems Research Committee," "Control of Water Quality in Transmission and Distribution Systems Committee," and "Emerging Technologies in Computers and Automation Committee." Dr. Goodrich is also active in the American Water Works Association Research Foundation serving on several past and current Project Advisory Committees. He is also an active member of the American Society of Civil Engineers and past member of the Task Committee on Risk and Reliability Analysis of Water Distribution Systems of the Committee on Probabilistic Approaches to Hydraulics of the Hydraulics Division.

CONTENTS

Point-of-use/ Point-of-entry

for Drinking Water Treatment

CHAPTER 1

THE SAFE DRINKING WATER ACT:
ITS REGULATORY IMPACT

INTRODUCTION

The Safe Drinking Water Act (SDWA) and its amendments have a dramatic impact on the way in which we view the treatment and distribution of water in the U.S. The water supply industry is undergoing a major revolution that may transform it beyond recognition. This revolution is being driven both by State and Federal regulations and by general advances in science and technology. Figure 1 illustrates some of the regulatory actions resulting from the Safe Drinking Water Act and its Amendments (1986). In addition to many new regulations, several "rules" are being considered to provide comprehensive guidance to the industry. These rules include the Surface Water Treatment Rule (SWTR) and the Disinfection and Disinfection By-Product Rule. Clearly, technology will play a major role in meeting these regulations including the introduction of ozonation, granular activated carbon, aeration, membrane processes, gas chromatography, mass spectrometry, and computerized process control. This leap to new levels of sophistication may be as important historically as the introduction of disinfection and filtration.[1]

Not only are the proposed changes comprehensive, the time frame for implementing these changes is extremely short. Their effect may well be to eliminate flexibility on the part of community and non-community systems in attempting to meet these regulations. Drinking water treatment has reached the point that changing one parameter or one treatment process may result in unintended or unexpected and perhaps undesirable effects on other parameters.

For example, chlorine has been used as a very effective disinfectant in drinking water since the turn of the century. However, it has been discovered that the interaction of chlorine and natural organic matter in water forms objectionable by-products. Many utilities and point-of-entry units use an alternate disinfectant such as ozone, but ozone will not maintain a disinfectant residual in the distribution system. Therefore, chlorination or chloramination will be required after ozonation. Another disinfectant being considered is chlorine dioxide which does not form

1

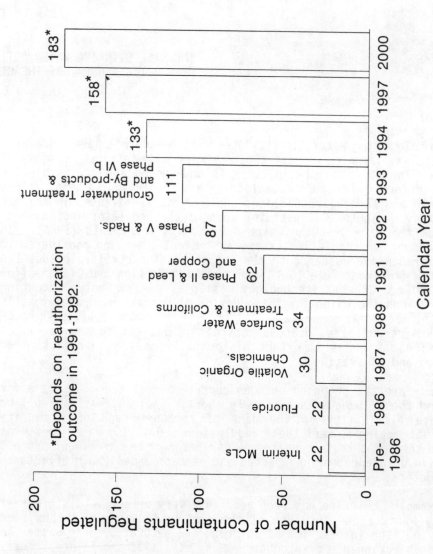

FIGURE 1. REGULATION HISTORY AND FUTURE SCHEDULE

some of the objectionable disinfection by-products but does form inorganic by-products that must be removed from treated water. Understanding the trade-off between minimizing objectionable by-products and maintaining bacterial safety is difficult. This is only one small example of some of the difficult problems facing the water supply industry today.[1]

Although the Safe Drinking Water Act is in a progressive regulatory development phase, once this phase is completed, a difficult period of implementation will lie ahead. As illustrated in Figure 1, the water supply industry will soon be inundated by waves of regulatory requirements. Meeting these regulations will require extensive information transfer and translation of research into practice. There are, however, many barriers to implementing the research knowledge currently available for meeting requirements of the Safe Drinking Water Act. A major barrier is the nature of water supply and water supply management itself. Although the provisions of the Safe Drinking Water Act and its amendments apply to public water systems, individual homeowners should strive to achieve the same quality of drinking water. Therefore, the discussion in this chapter is useful to both public and non-public drinking water systems and the approximately 20 million individuals drinking water from their own private wells.

NATURE OF THE WATER INDUSTRY

Safe Drinking Water Act regulations apply primarily to Public Water Supply systems. A Public Water Supply system is defined as a system which has at least fifteen service connections or regularly serves an average of at least twenty-five individuals daily at least 60 days out of the year. Public Water Supply systems are divided into Community Water Supply and Non-Transient, Non-Community Water Supply systems.[2]

A Community Water Supply system is a Public Water Supply system which serves year-round residents. A Non-Transient, Non-Community Water Supply system is a Public Water Supply system which serves primarily non-residential customers. Non-Community Water Supply systems include Transient systems (campgrounds, gas stations) and Non-Transient systems (schools, workplaces, hospitals) which have their own water supply and serve the same population for six months during a year. There are 201,242 water systems which meet the federal definition of Public Water Supply (PWS) systems. Twenty-nine percent (58,908) of the systems are Community Water Supply (CWS) systems serving primarily residential areas and 71% (142,334) are Non-Transient, Non-Community Water Supply (NTNCWS) systems serving primarily non-residential areas.

Figure 2 illustrates the categorization of Public Water Supply systems by type and source water. Surface water is the primary source of 19% of the CWS systems, which serve two-thirds of the nation. Ground water is used by 81% of the systems which serve one-third of the population. Of the Non-Transient, Non-Community Water Supply systems, 97% are served by ground water sources.

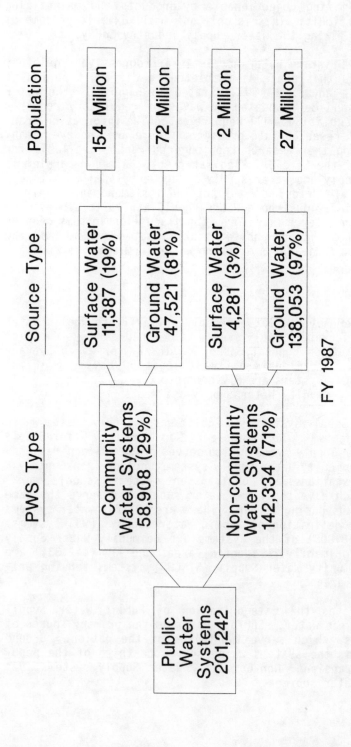

FIGURE 2. DISTRIBUTION OF PUBLIC WATER SYSTEMS BY
SYSTEM TYPE AND SOURCE WATER

All NTNCWS systems are small as are most Community Water Supply systems. Community Water Supply systems are usually categorized into five sizes (Table 1).

TABLE 1.   SIZE CATEGORIES FOR COMMUNITY WATER SUPPLY SYSTEMS

SIZE	POPULATION SERVED
Very Small	< 500
Small	501 - 3,300
Medium	3,301 - 10,000
Large	10,001 - 100,000
Very Large	> 100,000

Table 2 illustrates the distribution of Community Water Supply systems by size category. As can be seen, the vast majority of community water systems are very small but serve only 2.4% of the population.

TABLE 2.   COMMUNITY WATER SYSTEMS BY SIZE DISTRIBUTION

System Size (population served)	No. of CWS*	Percent of CWS	Population Served (million)	Percent Population Served
Very Small < 500	37,420	63.5%	5.5	2.4%
Small 501 - 3,300	14,132	24.0%	19.3	8.5%
Medium 3,301 - 10,000	4,203	7.1%	24.0	10.6%
Large 10,001 - 100,000	2,878	4.9%	78.5	34.8%
Very Large > 100,000	275	0.5%	98.7	43.7%
NATIONAL TOTAL	58,908	100%	226	100.0%

* Community Water Supplies

Economic Considerations

Implementation of the Safe Drinking Water Act will have a substantial economic impact on the industry. Consumers will, in general, bear the brunt of increased costs. Figure 3 estimates the cost to the nation of some of the rules and regulations shown in Figure 1.[3]

As illustrated by Figure 4, technology traditionally is assumed to exhibit significant economies of scale. Therefore, small utilities, even under the best of conditions, will pay higher unit costs for any type of technology than will large systems. Figure 5 shows the very high annual costs that may be borne by residents of small communities for both inorganic and organic control. Thus, point-of-use/point-of-entry drinking water treatment units may find a role in small communities and with individual homeowners in providing safe, aesthetically pleasing drinking water.

REGULATORY AGENDA

The 1986 Amendments to the SDWA require the United States Environmental Protection Agency (USEPA) to set Maximum Contaminant Level Goals (MCLGs) for many contaminants found in drinking water. These MCLGs must provide an adequate margin of safety from contaminant concentrations that are known or anticipated to induce adverse effects on human health. For each contaminant, the USEPA must establish a Maximum Contaminant Level (MCL) that is as close to the MCLG as is feasible with the use of Best Available Technology (BAT). Although the BAT identified for each contaminant must be an economically feasible and proven technology under field conditions, systems will not be required to install BAT for purposes of meeting a corresponding MCL.

If analytical techniques are not economically or technologically feasible for a given contaminant, then the USEPA must set a treatment technique for that contaminant in lieu of an MCL. The treatment technique must, in USEPA's judgement, be capable of providing economically feasible reduction of human health risks.

Treatment

The SDWA Amendments state that the USEPA must list the BATs capable of meeting MCL regulations. The SDWA Amendments also state that BAT for removing Synthetic Organic Chemicals (SOCs) must be at least as effective as Granular Activated Carbon (GAC). To date, BATs have been specified for the eight Volatile Organic Chemicals (VOCs) as shown in Table 3 and various SOCs, VOCs, and pesticides shown in Table 4.[4,5,6] Also shown are the potential health effects.

Technologies which have successfully removed VOCs and SOCs in site specific situations include: packed tower aeration (PTA), powdered activated carbon (PAC), diffused aeration, advanced oxidation processes, reverse osmosis (RO) and conventional treatment.

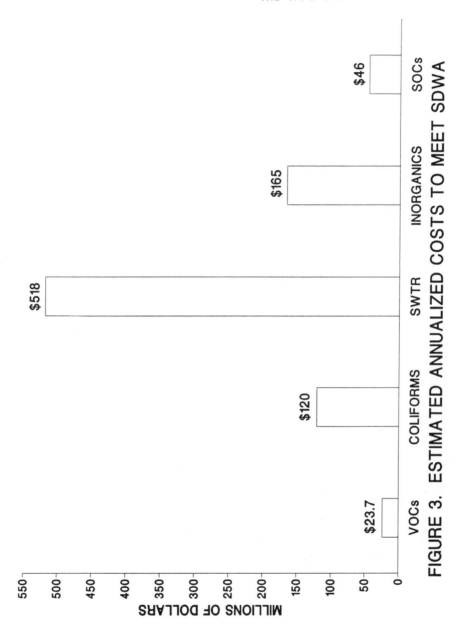

FIGURE 3. ESTIMATED ANNUALIZED COSTS TO MEET SDWA

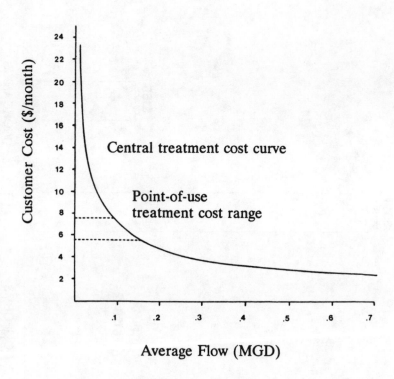

Figure 4. Central and Point-of-Use Treatment Costs for Activated Alumina

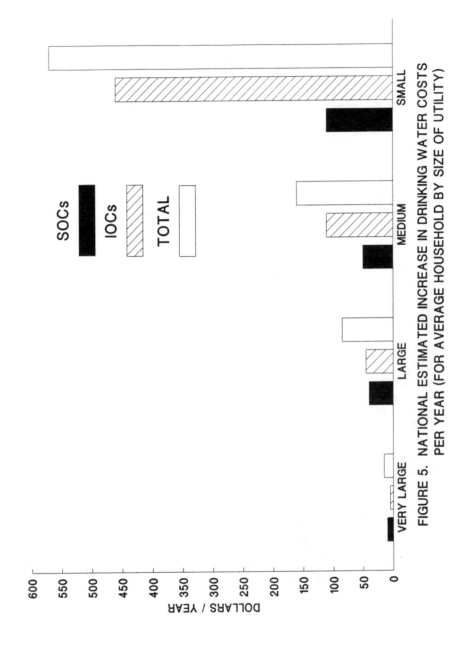

FIGURE 5. NATIONAL ESTIMATED INCREASE IN DRINKING WATER COSTS
PER YEAR (FOR AVERAGE HOUSEHOLD BY SIZE OF UTILITY)

TABLE 3.   FEDERAL REGULATIONS FOR VOLATILE SYNTHETIC
ORGANIC CHEMICALS

First Group VOCs·Final	MCLG (mg/L)	MCL (mg/L)	BAT[1]	CARCINOGEN CLASS[2]
Trichloroethylene	0	0.005	GAC, PTA	A
Carbon Tetrachloride	0	0.005	GAC, PTA	A
Vinyl Chloride	0	0.002	PTA	A
1,2-Dichloroethane	0	0.005	GAC, PTA	A
Benzene	0	0.005	GAC, PTA	A
para-Dichlorobenzene	0.075	0.075	GAC, PTA	B2
1,1-Dichloroethylene	0.007	0.007	GAC, PTA	B2
1,1,1-Trichloroethane	0.200	0.20	GAC, PTA	E

[1]   BAT = Best Available Technology
PTA = Packed Tower Aeration

[2]   Carcinogen classification of each chemical (as identified below
is based upon inhalation and ingestion data from short term and
chronic studies in humans and animals.   The key to carcinogen
classification is shown below:

- Group A - known human carcinogen
- Group B - probable human carcinogen
    * Group B2 - combination of adequate evidence in animals
                 and insufficient carcinogenic data in humans.
- Group E - no evidence of carcinogenicity in humans.   Possible
  liver, kidney and nervous system damage.

TABLE 4. PHASE II SOCs/VOCs/PESTICIDES

	MCLG (mg/L)	MCL (mg/L)	BAT[1]	DRINKING WATER HEALTH EFFECTS
Acrylamide	0	(2)	Limit Use	
Alachlor	0	0.002	GAC	probable cancer
Aldicarb	0.001	0.003	GAC	nervous system
Aldicarb sulfoxide	0.001	0.004	GAC	nervous system
Aldicarb sulfone	0.001	0.002	GAC	nervous system
Atrazine	0.003	0.003	GAC	reproductive and cardiac
Carbofuran	0.04	0.04	GAC	nervous system and reproductive
Chlordane	0	0.002	GAC	probable cancer
cis-1,2-Dichloroethylene	0.07	0.07	GAC, PTA	
Dibromochloropropane	0	0.002	GAC, PTA	probable cancer
1,2-Dichloropropane	0	0.005	GAC, PTA	
o-Dichlorobenzene	0.6	0.6	GAC, PTA	nervous system, lung, liver, kidney
2,4-D	0.07	0.07	GAC	liver, kidney, nervous system
Ethylenedibromide	0	0.00005	GAC, PTA	probable cancer
Epichlorohydrin	0	(2)	Limit Use	probable cancer liver, kidney, lungs
Ethylbenzene	0.7	0.7	GAC, PTA	kidney, liver, nervous system
Heptachlor	0	0.0004	GAC	probable cancer
Heptachlor epoxide	0	0.0002	GAC	
Lindane	0.0002	0.0002	GAC	nervous system, liver, kidney
Methoxychlor	0.04	0.04	GAC	nervous system, liver, kidney
Monochlorobenzene	0.1	0.1	GAC, PTA	kidney, liver nervous system
PCBs as decachlorbiphenyl	0	0.0005	GAC	
Pentachlorophenol	0	0.001	GAC	probable cancer, liver, kidney
Styrene	0.1	0.1	GAC, PTA	
Tetrachloroethylene	0	0.005	GAC, PTA	liver, nervous system
Toluene	1	1	GAC, PTA	kidney, nervous system, lungs
2,4,5-TP	0.05	0.05	GAC	nervous system, liver, kidney
Toxaphene	0	0.003	GAC	probable cancer
trans-1,2-Dichloroethylene	0.1	0.1	GAC, PTA	nervous system, liver, circulatory
Xylenes(total)	10	10	GAC, PTA	liver, kidney, nervous system

(1) GAC = Granular Activated Carbon
    PTA = Packed Tower Aeration

(2) Treatment Technique

The treatment and compliance methods available to a water utility searching for the most economical and effective means to comply with the proposed VOC/SOC MCLs include modification of existing treatment systems, installation of new systems, and the use of non-treatment alternatives, such as regionalization or alternate raw water sources. The major factors that must be considered in selecting a compliance method include:

- Quality and type of water source.

- Degree of SOC contamination.

- Specific contaminants present in source water.

- Economies of scale and the economic stability of the communities being served.

- Treatment and waste disposal requirements.

Some of the available methods are more complex than others. Selection of a technology by a community may require engineering studies and/or pilot-plant operations to determine the level of removal that any particular method will provide for that system. Similar requirements may be needed before installing POU/POE units.

Each of the new major regulatory activities will be discussed in the following section.

VOCs/SOCs

Typically, VOCs/SOCs are used in agricultural products, maintenance applications and industrial manufacturing processes. States have the option of setting more stringent regulations than the USEPA and can also set standards for additional contaminants.

A third group of MCLs and MCLGs were proposed in July of 1990 for VOCs and SOCs[7] (Table 5).

TABLE 5.   THIRD GROUP SOCs/VOCs (Proposed July 1990)

	PROPOSED MCLG (mg/L)	PROPOSED MCL (mg/L)	DRINKING WATER HEALTH EFFECTS
Dalapon	0.2	0.2	kidney, liver
Di(ethylhexyl)adipate	0.5	0.5	liver, reproductive system
Di(ethylhexyl)phthalate	0	0.004	
Dichloromethane	0	0.005	
Dinoseb	0.007	0.007	thyroid
Diquat	0.02	0.02	reproductive organs
Endothall	0.1	0.1	liver, kidney, GI tract, reproductive system
Endrin	0.002	0.002	liver, kidney, heart
Glyphosate	0.7	0.7	liver, kidney
Hexachlorobenzene	0	0.001	
Hexachlorocyclopentadiene	0.05	0.05	kidney, stomach
Oxamyl(Vydate)	0	0.2	kidney
PAHs(Benzo(a)pyrene)	0	0.0002	
Pichloran	0.5	0.5	kidney, liver
Simazine	0.001	0.001	circulatory system
1,2,4-Trichlorobenzene	0.009	0.009	kidney, liver
1,1,2-Trichloroethane	0.003	0.005	kidney, liver
2,3,7,8-TCDD	0	5×10^{-8}	reproductive, liver

The USEPA is also considering establishment of MCLGs and MCLs for six additional Polycyclic Aromatic Hydrocarbons (PAHs) classified as B2, probable human carcinogens shown in Table 6.

TABLE 6.   POSSIBLE PAH REGULATIONS

PAH	MCLG (mg/L)	MCL (mg/L)
Benz(a)anthracene	0	0.0001
Benzo(b)fluoranthene	0	0.0002
Benzo(k)fluoranthene	0	0.0002
Chrysene	0	0.0002
Dibenz(a,h)anthracene	0	0.0003
Indenopyrene	0	0.0004

LEAD/IOCs/RADs

In response to the problem of lead exposure, the USEPA established a regulation to cover both lead and copper corrosion.[8] Until recently, the Maximum Contaminant Level (MCL) for lead was 0.050 mg/L. The USEPA promulgated the MCL as an interim drinking water regulation in 1975. For copper, a secondary standard of 1 mg/L existed. On November 13, 1985, the USEPA began the process of revising the standards for lead and copper by proposing Maximum Contaminant Level Goals (MCLGs) for these two contaminants. However, the 1986 SDWA Amendments changed that plan. In addition to revising the drinking water standard for lead, the Safe Drinking Water Act (SDWA) includes other provisions that affect lead contamination of drinking water.

The 1986 SDWA Amendments banned the use of lead solder or flux (i.e., solder or flux containing more than 0.2 percent lead) and lead-containing pipes and fittings (i.e., pipes and fittings containing more than 9 percent lead). The lead ban became effective June 19, 1988. The SDWA Amendments also impose special public notification requirements regarding lead in drinking water.

Revised Standards

The Lead and Copper Rule, promulgated in June 1991, established MCLGs of zero for lead and 1.3 mg/L for copper. It also established an action level for treatment of 0.015 mg/L for lead and 1.3 mg/L for copper when these values are exceeded in 10% of household taps sampled.

Tap Water Monitoring for Lead and Copper

Monitoring for lead and copper is required every six months. One monitoring period is equivalent to six months; two monitoring periods are required per calendar year, i.e., January to June and July to December. Starting dates for monitoring are:

January 1992	Large systems (>50,000)
July 1992	Medium-sized systems (>3,300 to ≤50,000)
July 1993	Small systems (≤3,300)

Tap water samples are required from the following high risk locations:

-   homes with lead solder installed after 1982,
-   homes with lead pipes, and
-   homes with lead service lines.

Other IOCs and RADs

In addition to lead and copper, the USEPA promulgated regulations for nine inorganic contaminants (IOCs), as shown in Table 7.[6,7,8]

TABLE 7. FEDERAL REGULATIONS FOR INORGANIC CONTAMINANTS
(promulgated January 1991)

	MCLG (mg/L)	MCL (mg/L)	BAT[2]	DRINKING WATER HEALTH EFFECTS
Asbestos	7 MFL[1]	7 MFL[1]	C/F,DF,DMF,CC	Benign tumors
Barium	2	2	I/E,LS,RO,ED	Circulatory system
Cadmium[3]	0.005	0.005	C/F,LS,RO,IE	Kidney
Chromium	0.1	0.1	C/F,LS,RO,IE	Liver/kidney, skin and digestive system
Mercury	0.002	0.002	GAC,LS,C/F,RO	Kidney, nervous system
Nitrate	10	10	IE,RO,EDR	Methemoglobinemia "blue-baby syndrome"
Nitrite	1	1	IE,RO	Methemoglobinemia "blue-baby syndrome"
Total Nitrate/Nitrite	10	10	----	
Selenium	0.05	0.05	EDR,C/F,AA,LS,RO	Nervous system

1. MFL = million fibers per liter with fiber length > 10 microns

2. AA = Activated Alumina   CC = Corrosion Control
    C/F = Coagulation/Filtration   GAC = Granular Activated Carbon
    DF = Direct Filtration   LS = Lime Softening
    DMF = Diatomite Filtration   RO = Reverse Osmosis
    EDR = Electrodialysis Reversal

3. Promulgated June 1991[8]

Also, the USEPA proposed regulations for 6 additional inorganics (IOCs) in July 1990 and 4 radionuclides (RADs) plus 2 categories of radionuclides in June 1991; as shown in Table 8.[7,8]

TABLE 8.   PROPOSED FEDERAL REGULATIONS FOR IOCs (July 1990)
AND RADs (June 1991)

	PROPOSED MCLG (mg/L)	PROPOSED MCL (mg/L)	DRINKING WATER HEALTH EFFECTS
Inorganics			
Antimony	0.003	0.01/0.005	Decreases growth and longevity
Beryllium	0	0.001	Probable cancer, damage to bone and lungs
Cyanide	0.2	0.2	Spleen, brain, liver effects
Nickel	0.1	0.1	Heart, liver effects
Sulfate	400/500	400/500	Gastroenteritis
Thallium	0.0005	0.002/0.001	Kidney, liver, brain effects
	PROPOSED MCLG	PROPOSED MCL	DRINKING WATER HEALTH EFFECTS
Radionuclides			
Radium - 226	0	20 pCi/L	Human carcinogen
Radium - 228	0	20 pCi/L	Human carcinogen
Radon - 222	0	300 pCi/L	Human carcinogen
Uranium	0	20 ug/L	Human carcinogen
Adjusted gross alpha emitters	0	15 pCi/L	Human carcinogen
Gross beta and photon emitters	0	4 mrem ede/yr	Human carcinogen

SURFACE WATER TREATMENT RULE

The Surface Water Treatment Rule (SWTR) is the result of a series of amendments to the National Primary Drinking Water Regulations (NPDWR) and was promulgated on June 29, 1989:[9] Table 9 lists the MCL Goals contained in the rule.

TABLE 9.   MAXIMUM CONTAMINANT LEVEL GOALS

Giardia lamblia	0
Viruses	0
Legionella	0
Turbidity	none
Heterotrophic Plate Count	none

General Requirements

Coverage: All public water systems using any surface water or ground water under direct influence of surface water must disinfect, and may be required to filter, unless certain source water quality requirements and site specific conditions are met.

Treatment technique requirements are established in lieu of MCLs for *Giardia lamblia*, viruses, heterotrophic plate count bacteria, *Legionella* and turbidity.

Treatment must reliably achieve at least 99.9 percent (3-log) removal and/or inactivation of *Giardia lamblia* cysts, and 99.99 percent (4-log) removal and/or inactivation of viruses between the point where the raw water ceases to be subject to surface water runoff and a downstream point prior to delivery to the first customer.

All systems must be operated by qualified operators as determined by the State.

Compliance Dates

All systems using surface water or ground water under direct influence of surface water must meet the treatment requirements. Unfiltered systems must meet monitoring requirements within 18 months and meet criteria to avoid filtration beginning 30 months following promulgation. States must determine which systems must filter within 30 months after promulgation. Filtration must be installed within 18 months following failure to meet any one of the criteria to avoid filtration or 48 months following promulgation, whichever is later. Filtered systems must meet monitoring and performance requirements beginning 48 months following promulgation.

States must determine which community and non-community ground water systems are under direct influence of surface water within 5 years and 10 years, respectively, following promulgation. If a system is under direct influence of surface water, monitoring must begin 6 months following determination of direct influence. Filtration must be installed within 18 months following failure to meet any of the criteria to avoid filtration, or 48 months following promulgation, whichever is later.

SWTR Disinfection Requirements

Requirements for disinfection of surface waters are defined in the recently promulgated Surface Water Treatment Rule (SWTR). Under this rule, Ct values for *Giardia* cyst inactivation will meet disinfection requirements for unfiltered surface water supplies. For filtered supplies, disinfection requirements may be controlled by the Ct values for either *Giardia* cyst or virus inactivation, depending on site specific conditions.

Ct values are the product of the disinfectant residual concentration [mg/L] and contact time [minutes] measured at peak hourly flow. The concept of Ct was developed as a result of the SWTR, although it may also be applied for disinfection of ground water in the upcoming ground water disinfection rule.

Criteria Under Which Filtration is not Required

Unless the State determines that filtration is required, a water system that uses a surface water source or a ground water source under the direct influence of surface water is not required to provide filtration treatment if all of the following are met:

Samples must be taken immediately prior to the first point of disinfectant application after which the source is no longer subject to surface water runoff.

Coliform - Fecal coliform concentrations must be equal to or less than 20/100 or the total coliform concentration must be equal to or less than 100/100 ml in at least 90 percent of the measurements for the previous six months, calculated each month.

Minimum sampling frequencies for fecal or total coliform analysis are:

SYSTEM SIZE (persons)	Samples/Day
<501	1
501 - 3,000	2
3,301 -10,000	3
10,001 -25,000	4
> 25,000	5

If not already conducted during the minimum sampling frequency, coliforms must be analyzed for each day that the turbidity exceeds 1 NTU, unless the State determines that the samples cannot be analyzed, for logistical reasons outside the system's control, within 30 hours.

Turbidity - The turbidity level may not exceed 5 NTU. Filtration must be installed if 5 NTU is exceeded, unless the State determines that the exceedence does not occur for more than two periods in any consecutive 12 months, or five periods in any consecutive 120 months. An "event" is one or more consecutive days when at least one turbidity measurement each day exceeds 5 NTU. Turbidity levels must be measured every four hours by grab sample or continuous monitoring. States will establish the protocol for validating continuous monitoring.

Criteria for Filtered Systems

Turbidity Removal - Filtered water turbidity must be less than 5 NTU at all times.

Filtration Performance - Conventional filtration or direct filtration must achieve a filtered water turbidity equal to or less than 0.5 NTU in at least 95 percent of the measurements taken each month. The State may increase the turbidity limit to 1 NTU for at least 95 percent of the samples, without any demonstration by the system, if it determines the overall treatment, with disinfection achieves at least 99.9 percent and 99.99 percent removal/inactivation of Giardia cysts and viruses, respectively.

Disinfection For Water Systems

Disinfection - The inactivation achieved by disinfection must be at least 99.9 percent (3-log) inactivation of Giardia lamblia cysts and 99.99 percent (4-log) inactivation of viruses. Failure to meet this requirement on more than one day in a month is a violation. If a system is in violation for more than 1 of 12 months it must filter. Violation of this requirement may be allowed for 2 out of 12 months without requiring filtration if the State determines one failure to be caused by unusual and unpredictable circumstances. Each day of operation, the public water system must calculate the Ct value from the system's treatment parameters and determine whether this value(s) is sufficient to achieve the specified inactivation of Giardia lamblia cysts and viruses. For disinfectants other than chlorine a system may demonstrate, through the use of a State-approved protocol for on-site disinfection challenge studies or other information satisfactory to the State, that disinfection conditions other than the Ct values specified in the rule are adequate to achieve the specific inactivation rates. The following parameters must be monitored to calculate the inactivation based on Ct:

- Temperature once/day at each point of residual measurement.

- pH once/day at each point of residual measurements for chlorine.

- Disinfectant contact time daily at peak hourly flow.

- Residual disinfectant concentration(s) before or at the first customer each day during peak hourly flow.

Disinfection systems must have redundant components including alternate power supply with automatic start-up and alarms to ensure continuous disinfection of the water during operation, or have automatic shut-off of delivery of water to the distribution system whenever the disinfectant residual is less than 0.2 mg/L, provided that the State determines that a shut-off would not pose a potential health risk.

The system must demonstrate that the residual disinfectant concentration in the water delivered to the distribution system is never less than 0.2 mg/L for more than 4 hours. This some criteria should be used for POU/POE effluent. The system must notify the State any time the residual falls below 0.2 mg/L by the next business day. The notice must include the length of time the residual was less than 0.2 mg/L. If the residual is not restored within 4 hours, the system must filter unless the State determines the failure to be caused by unusual and unpredictable circumstances.

The residual entering the distribution system must be monitored continuously, and the lowest value recorded each day. Grab samples may be used during failure of continuous monitoring equipment for up to 5 days. Systems serving less than 3,300 may conduct grab sample monitoring in lieu of continuous monitoring at the following frequencies:

System Population	Samples/Day
≤ 500	1
501-1,000	2
1,001-2,500	3
2,501-3,300	4

Disinfection residuals in the distribution system must be detectable in at least 95 percent of the samples each month for any two consecutive months the system is in operation. Samples are to be taken at the same frequency as total coliforms. A system may measure for disinfectant residual or for heterotrophic bacteria. A heterotrophic plant count (HPC) concentration less than or equal to 500/mL is to be considered a positive residual concentration. For systems which cannot practically monitor for HPC or maintain a residual, the State may judge whether adequate disinfection is provided in the distribution system. A system must install filtration if it fails to meet the above requirement, unless the State determines the failure is not caused by a treatment deficiency of the source water. The SWTR list four disinfectants: Chlorine, chlorine dioxide, ozone and chloramines, and establishes "Ct" values corresponding to inactivation levels for each of these disinfectants.

For chlorine, ozone, and chloramines, the Ct for a 3-log *Giardia* cyst inactivation will provide more than a 4-log virus inactivation. For chloramines, the higher Ct for *Giardia* cyst inactivation is only true when chlorine is applied prior to ammonia. Unfiltered systems using these disinfectants should refer to the *Giardia* cyst Ct tables to determine whether the required inactivation levels are being met.

For chlorine dioxide, Ct values for 4-log virus inactivation may be higher than those for 3-log *Giardia* cyst inactivation for some temperature conditions. The system must therefore refer to the Ct tables for both *Giardia* cyst and virus inactivation for the appropriate temperature conditions to identify the controlling Cts for the system. The USEPA is continuing work to finalize the Ct values for chlorine dioxide.

Other Conditions

A system must maintain a watershed control program which will minimize the potential for contamination by human enteric viruses and *Giardia lamblia* cysts. Those using individual wells should be aware of this potential and protect their drinking water source. The State will determine the adequacy of the program to meet this goal. A system must monitor and control the activities in the watershed that may have an adverse impact on water. A system must demonstrate through ownership or written agreements with landowners in the watershed that it is able to limit and control all human activities that may have an adverse impact on water quality.

An annual "on-site inspection" to evaluate the adequacy of the watershed control program and the reliability of the disinfection treatment must be conducted by a party approved by the State with findings satisfactory to the State. A system must not have been identified as a source of a waterborne disease outbreak, or if it has been so identified, the system must have been modified sufficiently to prevent another such outbreak as determined by the State. Systems must not be out of compliance with the monthly MCL for total coliforms caused by a deficiency in treatment of the source water for any two months in any consecutive 12 month period. Systems serving more than 10,000 people must be in compliance with the MCL requirements for total trihalomethanes.

THE COLIFORM RULE

All public water systems must meet the requirements of the Coliform Rule, which sets an MCL requiring monitoring of total coliforms with frequency dependent on population served, and may call for a sanitary survey in small systems.[9]   Compliance is required 18 months after promulgation. The Coliform Rule is the result of a series of amendments to the National Primary Drinking Water Regulations (NPDWR).

Coliform Sampling

Total coliform samples must be taken at points representative of the distribution system, according to a written siting plan, approved by the State. Failure to meet the sampling requirements listed below must be reported to the State within 10 days and to the public within 14 days in the newspaper or within 45 days by mail.

Routine Sampling

   Community Water Systems - The required frequencies of total coliform
monitoring in community water systems depend on populations served. This
sampling must begin 18 months after the rule's promulgation. However, the
State may allow systems depending solely on ground water and serving fewer
than 1,000 people to reduce their monitoring frequency to once per quarter
if there is no history of total coliform contamination in the distribution
system and a sanitary survey finds no sanitary defects.

   Non-Community Water Systems - Non-community systems using all or part
surface water must monitor total coliforms at the same frequency as the
community systems.   Systems using ground water and serving fewer than
1,000 people must monitor quarterly, beginning 5 years after the Rule's
promulgation, although this can be reduced to yearly if a sanitary survey
shows no defects.

Repeat Sampling

   Any system that tests coliform-positive for a given sample must
collect a set of repeat samples within 24 hours of notification of the
positive result.   This time limitation may be relaxed on a case by case
basis, for certain systems at the State's discretion. The required number
of repeat samples will vary depending on the number of routine samples
taken.   Systems that collect up to one sample per month must take three
repeat samples. The location of these repeat samples must include the tap
where the original total coliform positive sample was collected, as well
as one tap upstream and one tap downstream, each within five service
connections of the first.

   When one or more repeat samples test total coliform-positive, an
additional set of repeat samples must be taken, as described previously.
This process continues until either no total coliform-positive samples
are found or the MCL is exceeded.   In systems collecting fewer than five
routine samples per month, if a total coliform-positive sample is taken,
the following month at least five routine samples must be collected,
except in certain cases where waived by the State.

Invalidation of Total Coliform Samples

   Total coliform-positive samples may be invalidated by the State under
a few conditions: first, where the laboratory finds positive results were
caused by improper analysis: second, if the State determines the total
coliform-positive results to be unrepresentative of the distribution system (in this case the State must document the reason for result invalidation); and third, if repeat samples show that the tap of the original
total coliform-positive sample produces all total coliform-positive
samples while all other taps produced total coliform-negative samples.
Samples may not however, be invalidated on the basis of total coliform-
negative repeat samples taken at the original tap.   Samples that are
invalidated for any reason are not included in the calculation of MCLs.

Sanitary Surveys

Any system collecting fewer than five samples per month on a regular basis must conduct sanitary surveys. Community and non-community systems must conduct the initial sanitary survey within 5 and 10 years, respectively. Subsequent sanitary surveys must be conducted every 5 years, except for non-community systems using protected and disinfected ground water, which have up to 10 years to do so. If the sanitary surveys are not conducted as specified, this non-compliance must be reported to the State and public in the same fashion described for sampling non-compliance.

Fecal coliforms/*Escherichia coli* (*E.coli*)

Any routine or repeat sample that tests coliform-positive must be analyzed for the presence of fecal coliform or *E. coli*. If either is present, the State must be notified within the next business day. In some cases the State may allow systems to assume total coliform-positive samples are fecal coliform-positive or *E. coli*. to avoid testing.

Maximum Contaminant Level (MCL)

To comply with the new Rule, systems must meet the MCL for total coliforms. This MCL is not based on a concentration, but rather on the presence or absence of coliforms in the samples. The MCL is calculated from all samples taken in a given month, whether they be routine or repeat samples. Under the MCL, no more than 50 percent of all total coliform samples per month can be total coliform-positive (systems collecting fewer than 40 samples per month can only have one total coliform-positive sample per month). When a system determines that it is out of compliance, it must report to the State within the next business day, and to the public within 14 days in the newspaper or 45 days in the mail.

As described previously, any sample that tests total coliform-positive must be analyzed for either fecal coliform or *E. coli*. Any fecal coliform or *E. coli* positive repeat sample, or any fecal coliform or *E. coli*. positive routine sample followed by a total coliform-positive routine sample followed by a total coliform-positive repeat sample constitutes non-compliance with the MCL. This non-compliance must be reported to the State within the next business day. Because it may pose an acute health risk, the public must be notified of this non-compliance via electronic media.

DISINFECTION AND DISINFECTION BY-PRODUCTS

New disinfection requirements will be established for all drinking water systems. Disinfection requirements for surface waters will be defined in the Surface Water Treatment Rule (SWTR). Ground waters, not affected by the SWTR, will probably be required to achieve a 99.99 percent inactivation/removal of enteric viruses.

Requirements for disinfection of ground waters are expected to be modeled after the SWTR. As a consequence, the Ct concept will ultimately apply to primary disinfection of ground waters as well as surface waters. However, while surface water disinfection will be controlled by Ct values for *Giardia* inactivation, primary disinfection practices for ground water treatment may be controlled by Ct values for inactivation of the Hepatitis A virus.

In addition to setting minimum required levels for disinfectants in distribution systems, the USEPA is also expected to set maximum contaminant levels (MCLs) for the same disinfectant residuals.[2]

Regulation of Disinfection By-Products (DBPs)

The USEPA has recently begun exploring options for regulation of DBPs. The major options being considered are:

Treatment Techniques - This strategy is applied when the contaminant of concern cannot be measured in a technologically or economically feasible manner. For DBP control, such an approach could limit disinfectant addition to only those waters containing less than a specified level of natural organic matter. This strategy could be used to limit the formation of ozone by-products such as formaldehyde, chloramine by-products such as N-organochloramines, and chlorine by-products such as MX [3-chloro-4-(dichloromethyl)-5-hydroxy-2(5H)-furanone].

MCLs for DBP Surrogates - This strategy could be applied for control of chlorine by-products by setting an MCL for surrogate parameters such as total organic halides (TOX).

MCLs for DBP Groups - Already being used to regulate total trihalomethanes (TTHMs, the sum of chloroform, bromodichloromethane, dibromochloromethane and bromoform concentrations), this concept may be extended to haloacetic acids, haloacetonitriles, chlorophenols and ketones.

MCLs for Individual DBPs - This approach could be used to regulate the THMs on an individual basis (i.e., one MCL for chloroform, one for bromoform, etc.). The USEPA is not expected to apply this strategy to DBPs which are conveniently measured in groups such as those noted in the above option. However, individual MCLs may be set for cyanogen chloride, chloropicrin and chloral hydrate.

The USEPA has not yet discussed the levels at which the MCLs may be set. The Safe Drinking Water Committee of the National Research Council (NRC) has recommended that the MCL for TTHMs be lowered below the present 100 ug/L. This has led to widespread speculation that this MCL will be reduced to at least 50 ug/L and possibly lower. However, the USEPA must set the MCLs at technologically and economically feasible levels. Economic feasibility may ultimately control the forthcoming MCLs since small water systems will also be required to meet the new requirements.

Anticipated Impacts

The impacts of these regulations are difficult to project at this time: The USEPA is evaluating the following issues:

Disinfection - Since the disinfection rules are expected to be similar in format to those specified by the SWTR, treatment facilities can anticipate similar impacts. Many systems may be required to install new or additional disinfection equipment and provide adequate actual (as opposed to theoretical) contact time in basins or clear wells.

Chlorine Dioxide - Use of chlorine dioxide is already quite limited by the existing non-enforceable guidance level. If this guidance level is upgraded to MCL and possibly lowered, chlorine dioxide may be limited to use as a primary disinfectant only in treatment systems that can demonstrate removal of chlorate and chlorite after their formation. Granular activated carbon (GAC) effectively removes chlorite but chemical reduction techniques (i.e., ferrous chloride) may be necessary for removal of both chlorine dioxide and chlorite. Chemical reduction techniques are being demonstrated for such a purpose.

Chloramines - MCLs can conceivably be set at levels that would preclude the use of chloramines in drinking water treatment. In addition, some utilities may need to use ammonia removal techniques if high levels of ammonia exist in the water source. The USEPA has also observed higher levels of cyanogen chloride in systems using chloramine treatment strategies. An MCL for cyanogen chloride or a treatment technique for N-organochloramine control could conceivably limit the use of chloramines as a DBP control strategy.

Ozone - Treatment techniques may limit the application of ozone to only those waters containing less than a specified level of natural organic matter. Such a regulation could prevent the use of ozone for certain applications, particularly pre-oxidation. Ozone has also been observed to increase the formation of DBPs such as chloropicrin and the haloketones when chlorine is utilized for secondary disinfection.

DBP Control - Developing a DBP control strategy requires careful planning. The current USEPA approach strongly favors those strategies involving removal of DBP precursors prior to disinfectant addition. These strategies may involve optimization of existing processes or addition of new processes for removal of natural organics. Many utilities have planned to approach DBP control by using alternate disinfectants such as ozone and chloramines.

While such an approach is excellent for control of THMs and TOX, it may not be adequate for control of other DBPs such as chloropicrin and the haloketones. Consequently, utilities that have already installed this approach for THM control will need to review the DBP issue once again. A limited number of utilities have installed processes, such as aeration, that remove THMs after their formation. These utilities will also need to reexamine their existing approach for control of the other DBPs.

Although these forthcoming regulations are difficult to anticipate, they will be a major driving force behind the design and operation of drinking water treatment facilities in the future and should be a consideration when selecting POU/POE treatment.

Strategies to Reduce Disinfection By-products

DBPs are produced or formed whenever a disinfectant or oxidant is added to water:

Natural organics + Disinfectant = DBP

Based on this relationship, three general strategies (or combinations thereof) are available for reducing DBPs in finished drinking water supplies:

1.   Remove the DBPs after they are formed.

2.   Use an alternate disinfectant/oxidant other than chlorine which does not produce undesirable DBPs.

3.   Remove the natural organics (precursors) before disinfection/oxidation.

Of these, the first two may be faulted for not treating the problem, but only dealing with symptoms. The third strategy gets to the root of the problem itself - natural organics or precursors found in all surface water supplies and, to a lesser degree, in many groundwater supplies.

Removal of DBPs

After DBPs are formed, several treatment technologies are available for removing them: oxidation, adsorption and aeration. However, each of these technologies has disadvantages regarding DBP removal.

Oxidation, even with ozone, is relatively ineffective for removing trihalomethanes (THMs), and can also form other non-THM DBPs. Adsorption, using granular activated carbon (GAC), is effective for removing DBPs but only for short periods of time, after which the GAC must be regenerated. Precursor materials are not easily stripped, and will still be present after aeration to form DBPs after a disinfectant/oxidant is added to the water. In addition, aeration cannot remove nonvolatile DBPs formed during treatment. In general, DBPs are very difficult, and, therefore, costly, to remove from drinking water once they have been formed.

Use of Alternative Disinfectant/Oxidant

Switching to alternative disinfectants/oxidants may be feasible provided the following criteria are met:

- DBPs are not produced at undesirable levels.

- Microbial inactivation is at least as effective as disinfection with chlorine.

- A stable residual is provided in the distribution system or home treated water.

From an economic standpoint, the alternative disinfectants/oxidants should be no more expensive than chlorine. Unfortunately, since no single existing alternative disinfectant/oxidant (e.g., ozone, chlorine dioxide, chloramines and UV radiation) can satisfy all of these requirements, a combination of alternative disinfectants/oxidants is needed.

For example, as an alternative disinfectant, a system might decide to use ozone, which is much more effective than chlorine for microbial inactivation and produces no significant halogenated organics unless chlorine is used as a secondary disinfectant. Because little is now known about ozonation by-products or their health effects, the USEPA is currently collecting these data and intends to regulate several ozone by-products.

Because ozone does not provide a stable residual for distribution system protection, chloramines might be used for secondary disinfection in combination with ozonation for primary disinfection. This may or may not be a problem in individual households.

While such a strategy is excellent for reducing THM and total halogenated organic by-product levels, the combined use of these disinfectants will produce other DBPs. Studies have shown that ozone/chloramine use can increase levels of chloropicrin, cyanogen chloride and total aldehydes. Consequently, systems using this strategy to reduce halogenated by-products, such as THMs, may have to reevaluate treatment as new DBP regulations are developed.

In summary, this strategy will most likely involve using more than one chemical to satisfy all requirements of an adequate alternative and will result in formation of other undesirable by-products, possibly requiring additional process modifications in the future.

Removal of Natural Organics/Precursors

Removal of natural organics, or precursor materials, prior to disinfection represents an optimum approach for controlling DBPs. Because precursor material is a constituent of the dissolved organic carbon (DOC) in raw water, optimizing treatment to remove DOC before adding a disinfectant/oxidant will provide the best strategy for reducing DBPs.

Treatment Techniques

    Available treatment technologies include: conventional treatment, oxidation, adsorption, and membrane processes.

    <u>Conventional Treatment</u> - Conventional treatment (coagulation/floc-culation/sedimentation/filtration) can be optimized for DBP precursor removal by converting the dissolved organics in raw water to particulate matter which is then removed during sedimentation and filtration. Some approaches are:

- Adjust coagulant dosage

- Reduce pH of coagulation

- Use another coagulant

- Use polymers as coagulant and/or filter aids

- Add powdered activated carbon (PAC)

- Adjust chemical mixing conditions

    <u>Oxidation</u> - Once existing conventional processes are optimized, an oxidant might be added in the early stages of treatment to assist in organics removal. Ozone, hydrogen peroxide, potassium permanganate, or combinations of these might be applied to the water during one or more of the treatment steps ahead of filtration to oxidize the organics and/or aid in flocculation. Recent bench-scale and pilot studies have indicated that, in some waters, ozone and a combination of ozone and hydrogen pero-xide result in increased DOC removals. In other waters they had no effect or resulted in an increase in THM formation potential. Oxidation can also increase the precursor concentration of some DBPS, and thus must be thor-oughly tested prior to implementation as a precursor control strategy.

    <u>Adsorption</u> - After maximum removal of dissolved organics via conven-tional treatment, if this is feasible, GAC may be applied either in com-bination with filtration or in a post filtration mode to achieve addi-tional removal. GAC will also remove many synthetic organic chemicals present in the water supply.

    <u>Membranes</u> - Membranes are an emerging technology for DBP precursor removal, with demonstrated applicability under relatively low pressure conditions and at high recovery rates. Bench-scale testing of several low pressure membranes has shown THM precursor removals of 90% and higher, although in actual practice, it may be difficult to achieve greater than 80 to 85% recovery of product water. Membrane technology, often used to

soften Florida ground waters, has also achieved high levels of THM precursor removal from waters containing significant concentrations of both hardness and dissolved organics. Many Florida water supplies are currently investigating membrane systems for both softening and DBP control.

Other Disinfection By-products

The USEPA released a short list of "ozonation" by-products during the April 6, 1990 meeting of the Drinking Water Subcommittee of the Science Advisory Board. Combining this list with the "chlorination" by-product list gives the list of DBPs (Table 10) that will receive focused attention for USEPA research on health effects, analytical methods, occurrence and treatment methods.[2]

Ozone can produce organic oxidation by-products, brominated halogenation by-products and inorganic by-products. However, it is important to note that all oxidants and disinfectants can produce many of the listed by-products to some extent.

The occurrence of different DBPs should be viewed in light of the Surface Water Treatment Rule (SWTR) requirements for primary and secondary disinfection, which essentially limit disinfection practices to these strategies:

- Chlorine/Chlorine
- Chlorine/Chloramine
- Ozone/Chlorine
- Ozone/Chloramine
- Chlorine Dioxide/Chlorine
- Chlorine Dioxide/Chloramine

All disinfection strategies are capable of producing many of the listed by-products. For example, ozone has also been observed to increase the formation of DBPs such as chloropicrin and the haloketones when chlorine is utilized for secondary disinfection.   In addition, preoxidation with chemicals such as potassium permanganate and hydrogen peroxide will also impact the by-product levels formed by the above disinfection strategies.

TABLE 10.   USEPA's SHORT-LIST OF DBPs

Organic Halogenation By-products
Total Trihalomethanes
  Chloroform
  Bromodichloromethane
  Dibromochloromethane
  Bromoform

Total Haloacetic Acids
  Monochloroacetic Acid
  Dichloroacetic Acid
  Trichloroacetic Acid
  Monobromoacetic Acid
  Dibromoacetic Acid

Total Haloacetonitriles
  Trichlroacetonitrile
  Dichloroacetonitrile
  Bromochloroacetonitrile
  Dibromoacetonitrile

Total Haloketones
  1,1-Dichloropropanone
  1,1,1-Trichloropropanone

Total Chlorophenols
  2-Chlorophenol
  2,4-Dichlorophenol
  2,4,6-Trichlorophenol

Chloropicrin
Chloral Hydrate
Cyanogen Chloride
N-Organochloramines
MX[1]

Disinfectant Residuals
Free Chlorine
  Hypochlorous Acid
  Hypochlorite Ion

Combined Chlorine (Chloramines)
  Monochloramine
  Dichloramine
  Trichloramine

Chlorine Dioxide

Inorganic By-products
  Chlorate
  Chlorite
  Bromate
  Iodate
  Hydrogen Peroxide

Ozone Oxidation By-products

Total Aldehydes
  Formaldehyde
  Acetaldehyde
  Hexanal
  Heptanal

Total Carboxylic Acids
  Hexanoic Acid
  Heptanoic Acid

Assimilable Organic Carbon

---

[1] [3-Chloro-4-(dichloromethyl)-
       5-hydroxy-2(5H)-foranone]

UNREGULATED CONTAMINANTS

A rule published in the July 8, 1987 issue of the Federal Register requires monitoring of 51 unregulated contaminants, which are divided into three groups.[4]   All systems are required to monitor every five years for the first group of 34 contaminants. Vulnerable systems must monitor for an additional group of two contaminants, while monitoring of the third group of remaining contaminants is at the discretion of the State.

The May 22, 1989 issue of the Federal Register proposed an additional 113 unregulated contaminants, divided into two groups.[9]   Monitoring of the first group of contaminants, including 23 organics and 6 inorganics, will only be required for vulnerable systems. Monitoring of the remaining 84 contaminants will be at the State's discretion.

BEST AVAILABLE TECHNOLOGY

It is acceptable to use any technology approved by the State as long as all of the water is treated and complies with the MCLs. There are many types of technologies which achieve the equivalent performance to the recommended BAT. The USEPA proposed in November, 1985 that point-of-use (POU) and point-of entry (POE) technologies should not be considered Best Technology Generally Available (BTGA), but would be considered acceptable for meeting MCLs under certain conditions. A major concern regarding the use of POU and POE technology is the problem of monitoring treatment performance so that it is comparable to central treatment. Of course, POU devices only treat water at an individual tap and therefore raise the possibility of potential exposure at untreated taps and do not treat contaminants introduced by showers and dermal contact. These devices are also not generally assumed to be affordable by large metropolitan water systems which is one of the criteria for setting BAT.

ROLE OF POU/POE

Commentaries on the November notice reflected a disagreement as to whether POU/POE devices should be considered BAT. Most of the commentaries agreed that POU/POE devices should not be considered as BAT. This general concern related to the problem associated with controlling installation, maintenance, operation, repair and potential human exposure resulting from untreated taps. Other commentaries agreed that POU/POE devices should be considered BAT or allowed for compliance, because these technologies may be more cost effective for small systems than central treatment.

In the final rule, POU and POE devices are not designated as BAT because: (1) POU and POE devices are difficult to monitor in a manner that assures the performance is equivalent to central treatment; (2) these devices are generally not affordable by large metropolitan water systems; (3) POU devices are not considered acceptable means of compliance with MCLs because they do not treat all the water in the house and could result in health risk due to exposure to untreated water. Therefore POU devices were considered acceptable for use as interim measures such as a condition of obtaining a variance or exemption to avoid unreasonable risks to health before full compliance can be achieved.

Under this rule POE devices are considered an acceptable means of compliance because POE can provide drinking water that meets MCLs at all points in the home. It is also possible that POE devices may be cost effective for small systems or non-transient non-community water systems. In many cases these devices are essentially the same as central treatment. It is recognized that operational problems may be greater for POE than for central treatment in a community system.

The USEPA has therefore imposed the following conditions on systems that use POE for compliance.

(1) In order to ensure that the device is kept in working order, the public water system is responsible for operating and maintaining all parts of the treatment system although central ownership is not necessary.

(2) An effective monitoring plan must be developed and approved by the State before POE devices are installed. A unique monitoring plan must be installed that ensures that the POE device provides health protection equivalent to central water treatment. This means this device must meet all Primary and Secondary Drinking Water Standards. In addition to meeting MCLs, monitoring measurements must include information on total flow treated and the mechanical conditions of the treatment equipment among other measurements.

(3) Because there are no generally accepted standards for design and construction of POE devices and there are a variety of designs available, the State must require adequate certification of performance and field testing. A rigorous engineering design review of each type of device is required. Either the State or a third party acceptable to the State can conduct a certification program.

(4) A key factor in applying POE treatment is the maintenance of the microbiological safety of treated water. There is a tendency for POE devices to increase bacterial concentrations in treated water. This is a particular problem for activated carbon technologies. Therefore, it may be necessary to use frequent backwashing, post-contactor disinfection, and monitoring to ensure the microbiological safety of the treated water. The USEPA considers this a necessary condition because disinfection is not normally provided after point-of-entry treatment as is commonly used in central treatment.

(5) The USEPA requires that every building connected to a public
     water system must have a POE device that is installed,
     maintained and adequately monitored.   The rights and
     responsibilities of the utility customer must be transferred to
     the new owner with the title when the building is sold.

REFERENCES

1. Clark, R. M., "Translating Research into Practice: The Drinking
    Water Industry", Proceedings of the Environmental Quality and
    Industrial Competitiveness Workshop, American Academy of
    Environmental Engineers, Baltimore, MD., April 11, 1989,

2. Clark, S. W., "Design and Use of Granular Activated Carbon -
    Practical Aspects", Proceedings of a Technology Transfer Conference
    by AWWARF and the U.S. Environmental Protection Agency, Cincinnati,
    Ohio pp. 1-14., May 9-10, 1989,

3. "Estimates of the Total Benefits and Total Costs Associated with
    Implementing the 1986 Amendments to the SDWA",   U.S.EPA, November
    1989.

4. "Synthetic   Organic   Chemicals;   Monitoring   for   Unregulated
    Contaminants: Final Rule", National Primary and Secondary Drinking
    Water Regulations, *Federal Register* 52(130):25701 (1987).

5. National Primary Drinking Water Regulations: Final Rule, *Federal
    Register*, 56(20):3526 (1991).

6. National Primary Drinking Water Regulations: Final Rule, Federal
    Register, 56(126):30266 (1991).

7. Synthetic Organic Chemicals and Inorganic Chemicals: Proposed Rule",
    National Primary and Secondary Drinking Water Regulations, *Federal
    Register*, 55(143):30370 (1990).

8. Maximum Contaminant Level Goals and National Primary Drinking Water
    Regulations for Lead and Cooper: Final Rule, *Federal Register*,
    56(110):26461 (1991).

9. "Filtration, Disinfection, Turbidity, Giardia lamblia, Viruses,
    Legionella, and Heterotrophic Bacteria: Final Rule", Drinking Water;
    National Primary Drinking Water Regulations, *Federal Register*,
    54(124):27486 (1989).

CHAPTER 2

INTRODUCTION

The selection of an individual or combination of devices for removing contaminants from a drinking water source is usually based on a specific requirement. In this selection process, one should perform a systematic technical evaluation of the available treatment processes along with an economic assessment. When evaluating various POU/POE treatment systems, six major factors should be considered in the decision process.[1] These factors include: (1) quality and type of water source, (2) type and extent of contamination, (3) economies of scale and cost of water, (4) treatment requirements (5) any waste disposal requirements, and (6) institutional requirements. Consultation with the State agency responsible for public water supplies and public or private water quality professionals is recommended to ensure that appropriate treatment technology and equipment is selected. In the following discussion, available POU/POE treatment devices for removal of various contaminants are presented.

TREATMENT TECHNOLOGIES

Currently, POU/POE treatment is used to control a wide variety of contaminants in drinking water. Basically, the same technology used in treatment plants for community water systems can be used in POU/POE treatment. This technology is applied to reduce levels of organic contaminants, control turbidity, fluoride, iron, radium, chlorine, arsenic, nitrate, ammonia, microorganisms including cysts, and many other contaminants. Aesthetic parameters such as taste, odor, or color can also be improved with POU/POE treatment.

Types of treatment include adsorption, ion-exchange, reverse osmosis, filtration, chemical oxidation, distillation, aeration, and disinfection such as chlorination, ozonation, and ultraviolet light. A list of contaminants and appropriate treatment technologies for removing some of these contaminants from drinking water is shown in Tables 1 and 2.[2,3]

TABLE 1.   PERFORMANCE SUMMARY FOR ORGANIC TREATMENT TECHNOLOGIES[2]

Organic Compounds	Regulatory phase	Removal efficiency				
		Granular activated carbon[a]	Packed-tower aeration	Reverse osmosis	Ozone oxidation (2-6 ppm)	Conventional treatment
VOCs						
Alkanes						
Carbon tetrachloride	I	++	++	++	-	-
1,2-Dichloroethane	I	++	++	+	-	-
1,1,1-Trichloroethane	I	++	++	++	-	-
1,2-Dichloropropane	II	++	++	++	-	-
Ethylene dibromide	II	++	++	+	-	-
Dibromochloropropane	II	++	+	NA	-	-
Alkenes						
Vinyl chloride	I	+	++	NA	++	-
1,1-Dichloroethylene	I	++	++	NA	++	-
cis-1,2-Dichloroethylene	II	++	++	-	++	-
trans-1,2-Dichloroethylene	II	++	++	NA	++	-
Trichloroethylene	I	++	++	+	+	-
Aromatics						
Benzene	I	++	++	-	++	-
Toluene	II	++	++	NA	++	-
Xylenes	II	++	++	NA	++	-
Ethylbenzene	II	++	++	-	++	-
Chlorobenzene	II	++	++	++	+	-
o-Dichlorobenzene	II	++	++	+	+	-
p-Dichlorobenzene	I	++	++	NA	+	-
Styrene	II	++	++	NA	++	-
PESTICIDES						
Pentachlorophenol	II	++	-	NA	++	NA
2,4-D	II	++	-	NA	+	-
Alachlor	II	++	++	++	++	-
Aldicarb	II	++	-	++	NA	-
Carbofuran	II	++	-	++	++	-
Lindane	II	++	-	NA	-	-
Toxaphene	II	++	++	NA	NA	-
Heptachlor	II	++	++	NA	+	NA
Chlordane	II	++	-	NA	NA	NA
2,4,5-TP	II	++	NA	NA	+	NA
Methoxychlor	II	++	NA	NA	NA	NA
OTHER						
Acrylamide	II	NA	-	NA	NA	NA
Epichlorohydrin	II	NA	-	NA	-	NA
PCBs	II	++	++	NA	NA	NA

++ =   Excellent (70-100%).
 + =   Average (30-69%)
 - =   Poor (0-29%)
NA =   Data not available.

TABLE 2.   PERFORMANCE SUMMARY FOR INORGANIC TREATMENT TECHNOLOGIES[3]

	Activated Alumina	Coagulation/ Filtration	Corrosion Control	Direct Filtration	Diatomite Filtration	Granular Activated Carbon	Ion Exchange	Lime Softening	Reverse Osmosis	Air Stripping
				REMOVAL EFFICIENCY						
Asbestos	-	++	++	++	++	-	-	-	-	-
Barium	-	-	-	-	-	-	++	++	++	-
Cadmium	-	++	-	-	-	-	++	++	++	-
Chromium III	-	++	-	-	-	-	++	++	++	-
Chromium IV	-	++	-	-	-	-	++	-	++	-
Mercury	-	+to++	-	-	-	++	-	++	++	-
Nitrate and Nitrite	-	-	-	-	-	-	++	-	++	-
Selenium IV (Senenite)	++	++	-	-	-	-	-	+	++	-
Selenium VI (Selenate)	++	-	-	-	-	-	-	-	++	-
Arsenic III	++[a]	++[a]	-	-	-	-	-	+to++[a]	+	-
Arsenic V	++	++	-	-	-	-	++	+to++	++	-
Radium 226	-	-	-	-	-	-	++	++	++	-
Radon	-	-	-	-	-	++	-	-	-	++
Uranium	[b]	++	-	-	-	-	++	++	++	-

++ = Excellent (70 - 100%)
+ = Average (30 - 69%)
- = Poor (0 - 29%)
a = with preoxidation
b = unknown

There are other constituents in drinking water that oftentimes should be removed although they may not be considered a health problem and may not regulated. These consist of things such as total dissolved solids, copper, chloride, sulfate, iron, manganese, color, taste, and odor. Basic treatment techniques such as reverse osmosis, ion exchange, and activated carbon can be used to remove these contaminants from drinking water.[4]

Activated Carbon

Activated carbon is the most widely used process for POU/POE systems for home treatment of water. Carbon units are normally the easiest to install and maintain. Operating costs are usually limited to filter replacement and their performance is sufficient for removing organic and some inorganic contaminants. The performance of an individual unit depends on a combination of factors, such as: (1) unit design, (2) type and amount of activated carbon, and (3) contact time of the water with the carbon.[5] Most units use granular activated carbon in their design, although other types of carbon such as pressed block, briquettes of powdered carbon are available.

Activated carbon removes contaminants from water by adsorption or the attraction and accumulation of one substance on the surface of another. In general, high surface area and pore structure are the prime considerations in adsorption of organics from water; whereas, the chemical nature of the carbon surface is of relatively minor significance.[6] Activated carbon has a preference for organic compounds and because of this selectivity is particularly effective in removing these compounds from water. The ability of activated carbon to adsorb large quantities of material is directly related to its porous nature. Much of the surface area available for adsorption in granular carbon particles is found in the pores within the granular carbon particles created during the activation process (Figure 1).

Various pore size ranges are created and have been described as shown below.[7]

Micropores		$r_p < 1nm$
Mesopores	$1nm <$	$r_p < 25nm$
Macropores	$25nm <$	r_p

where r_p is the pore radius

The smaller the pores with respect to the molecules to be adsorbed, the greater the forces of attraction. In the macropores, the specific surface area is small and contributes very little to adsorption.[7] The most common raw materials used in the production of the activated carbons suited for water treatment are bituminous coals, peat, lignite, petrol coke, wood, and coconut shells. Without extensive pretreatment, each raw material tends to produce a unique pore structure as shown in Figure 2.

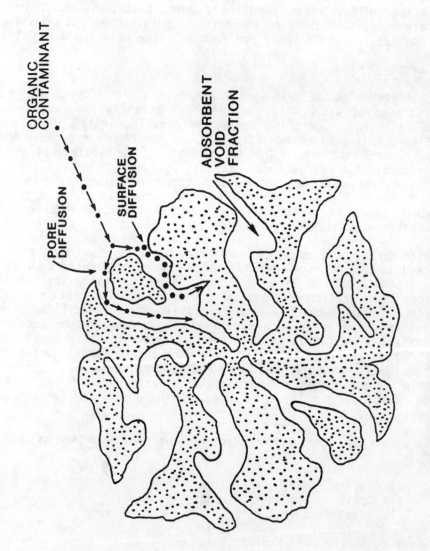

FIGURE 1. TYPICAL CARBON PORE

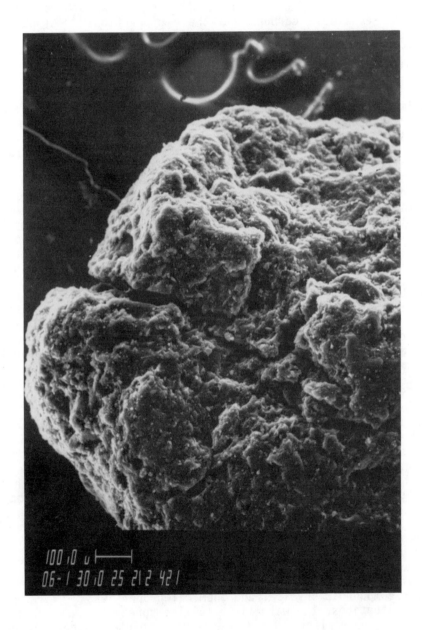

Figure 2.   Scanning Electron Micrograph of Filtrasorb 400 GAC Particle
(200X Magnification)

Each commercially available granular activated carbon has properties making it more suitable for certain applications than others.   Table 3 lists typical available carbons and their properties.[8]   A number of granular activated carbons are commercially available and the type most suited for a given application should be determined by laboratory testing based on isotherms.   Mathematical models have been developed to predict adsorption behavior in carbon systems, although not always reliably.

Two important variables when considering a carbon system are contact time and breakthrough.   These variables, along with flow quantity, determine what the volume should be of the carbon bed.   Carbon exhaustion is related to the breakthrough characteristics.   Breakthrough occurs when the effluent concentration after passing through carbon exceeds a predetermined performance value.   Once breakthrough occurs, the carbon has to be replaced.   Figure 3 shows a typical breakthrough curve and relates it to the relative amount of the carbon bed exhausted to the adsorption zone that is still available to adsorb contaminants.   At startup, water emerging from a carbon system contains very low or no concentration of contaminants.   As more liquid flows through the carbon, the adsorption capacity decreases and a gradual increase in effluent water contaminant concentration occurs.   As the carbon nears exhaustion, the water effluent contaminant concentration increases rapidly as it approaches the influent concentration.

The volume of activated carbon and the quantity of flow as well as the water quality are important in determining breakthrough, exhaustion, and the time to carbon replacement.   Generally, the longer the water is in contact with the carbon, the longer the time until breakthrough as shown in Figure 4.

Although activated carbon is considered a broad spectrum adsorbent, it does have a finite capacity for any one compound.[9]   When multiple compounds occur in drinking water, they compete for available sites on the carbon, reducing the capacity for the less strongly adsorbed compounds.   Therefore, the desired performance cannot be achieved without an activated carbon that has a significant capacity for the chemicals of concern.[9]

As mentioned earlier, isotherms, which represent the ultimate capacity of a compound on carbon, can be used to get an indication of the effectiveness of activated carbon in treating contaminated waters.   Also, a number of adsorption models have been proposed to predict isotherms by knowing the compound's physical and chemical properties.[10,11,12]   In most real water situations, the water is a multicomponent mixture of contaminants as discussed earlier.   Therefore, single-solute isotherms should be used with caution because of their limitations of predicting a capacity that is higher than the true capacity which is influenced by adsorptive competition.[13]   In order to incorporate the competitive effects into predicting the capacity of activated carbons, a number of multicomponent

TABLE 3. PROPERTIES OF TYPICAL GRANULAR ACTIVATED CARBONS

	TYPE A	TYPE B	TYPE C	TYPE D
Physical Properties				
Surface area, m^3/g (BET)	600-630	950-1050	1000	1050
Apparent density, g/cm^3	0.43	0.48	0.48	0.48
Density, backwashed and drained, lb/ft^3	22	26	26	30
Real density, g/cm^3	--	2.1	2.1	2.1
Particle density, g/cm^3	2.0	1.3-1.4	1.4	0.92
Effective size, mm	1.4-1.5	0.8-0.9	0.85-1.05	0.89
Uniformity coefficient	0.8-0.9	1.9	1.8	1.44
Pore volume, cm^3/g	1.7	0.85	0.85	0.60
Mean particle diameter, mm	1.6	1.5-1.7	1.5-1.7	1.2
Specifications				
Sieve size, U.S. std. series				
Larger than No. 8 max. percentage	8	8	8	--
Larger than No. 12 max. percentage	--	--	--	5
Smaller than No. 30 max. percentage	5	5	5	--
Smaller than No. 40 max. percentage	--	--	--	5
Iodine No.	650	900	950	1000
Abrasion No., minimum	--	70	70	85
Ash, percentage	--	8	7.5	0.5
Moisture as packed, max. percentage	--	2	2	1

-- = Data not available.

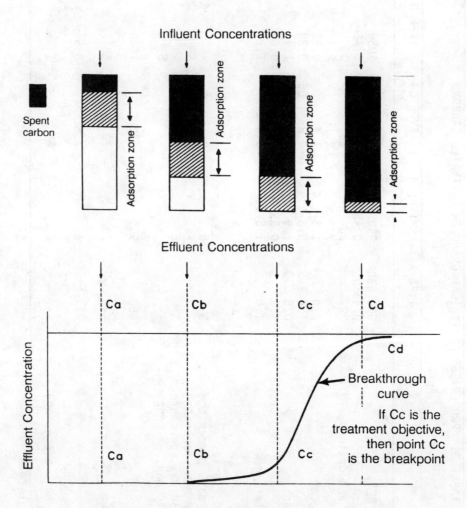

Figure 3. Passage of a Contaminant Through a Carbon Bed and
Corresponding Breakthrough Curve

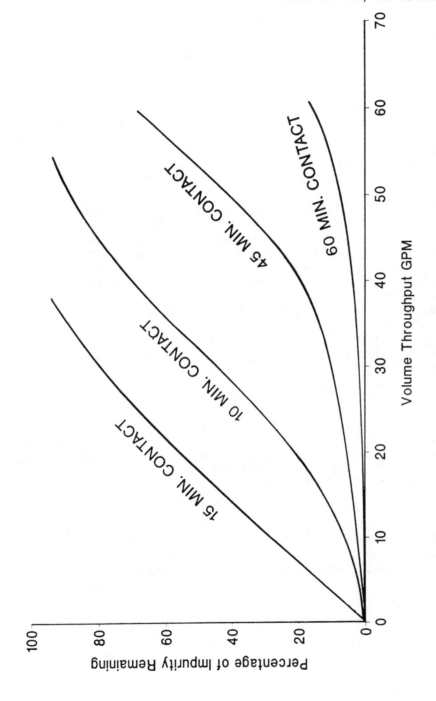

Figure 4. Typical Carbon Adsorption Breakthrough Curves
Showing Effect of Contact Time

models have been developed.[14,15,16,17,18] The authors suggest that when the consumer selects an activated carbon system they consider the limitations presented above and not assume that all activated carbon systems perform the same.

Activated carbon also provides an ideal medium for the accumulation and growth of living organisms. Concern, therefore, arises that direct consumption of water from POU/POE carbon devices may cause a bacterial health problem while preventing a chemical health problem (see Chapter 5). A typical POU activated carbon unit for removing organics from drinking water is shown in Figure 5.

Recently, there has been a concern about radon in water entering the home. Part of this concern comes from an estimated additional 100 to 1,800 lung cancer deaths per year caused by inhaling radon emitted by household water. Granular activated carbon is one method for removing radon. It was found to be effective because of an adsorption-decay that is established and continues for years.[19] However, the granular activated carbon adsorbs radon and as it decays gamma radiation is produced. The amount of radiation is related to the level of radon and other radioactive materials in the water supply and the amount of water used. To safely use granular activated carbon for removing radon, shielding may be required to lower radiation emissions to acceptable levels (See Chapter 4). Typical removals are 90 to 95 percent.

Membranes

A membrane is usually described as a thin, soft, pliable polymer material that forms a limiting surface or interface which governs the selection and passage of contaminants in the case of water treatment. Four specific polymer classes constitute the majority of membrane elements manufactured that are applied to drinking water treatment. They consist of polyamide, thin film composite, cellulose acetate, and cellulose triacetate.[20] These membrane polymers are useable only after they have been packaged into an element or module that can be efficiently and effectively used. The ideal membrane configuration has a high ratio of membrane area to volume and provides the greatest resistance to fouling from suspended solids. When selecting a membrane polymer for a specific application, one has to be careful because various polymer classes perform differently as shown in Table 4.

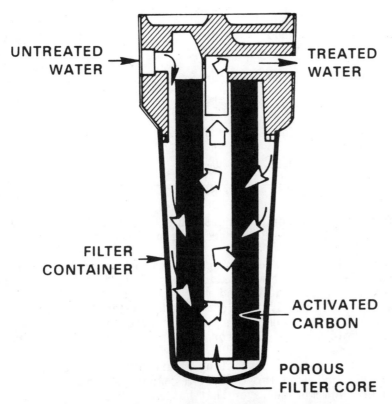

UNTREATED WATER

TREATED WATER

FILTER CONTAINER

ACTIVATED CARBON

POROUS FILTER CORE

(WATER QUALITY ASSOCIATION®)

FIGURE 5.   CROSS-SECTIONAL VIEW OF A TYPICAL ACTIVATED CARBON UNIT

TABLE 4.   COMPARISON OF MEMBRANE POLYMER PERFORMANCE[20,21,22,23]

| Parameters | Membrane Polymer Type | | | |
	Polyamide	Cellulose Acetate	Cellulose Triacetate	Thin Film Composite
pH tolerance	4-11	2-8	4-8	2-12
Chlorine tolerance	poor	good	fair-good	good
Biological resistance	good	poor	fair-good	good
Foul potential	moderate	low	low	moderate/high
Cleanability	good	poor-fair	poor-fair	good
Temperature limit for stability, °C	35	35	30	50
Typical rejection of ionic species, %	>90	90	90	>90
Typical rejection of organic compounds, %	0-90	0-35	--	0-100

Before proceeding further, one needs to understand the terms commonly used in membrane technology. Below are some of these terms.

Cellulose Acetate and Triacetate - An acetic acid ester of cellulose which is a tough, flexible, thermoplastic material used in the manufacture of synthetic ultrafiltration and reverse osmosis membranes.

Thin-Film Composite - Thin synthetic membrane bonded to an anisotropic support layer. This composite is supported by a third layer of synthetic fabric and is used for both ultrafiltration and reverse osmosis membranes.

Polyamide - A membrane polymer consisting of aromatic diamide and aromatic acid chlorides and used for thin film composite construction.

Anisotropic Layer - A membrane layer that has an increase in porosity from the surface to the base.

Polysulfone Resin - A thermoplastic polymer containing sulfone linkage with excellent high-temperature, low creep, and arc resistance properties. This polymer is used as a primary material in the manufacture of membranes or as a backing for thin-film composite membrane material.

Reverse Osmosis - A membrane separation process which uses pressure above osmotic pressure to force water through a semipermeable membrane that rejects dissolved contaminants while allowing water to pass through the membrane.

Ultrafiltration - Separation of colloidal or dissolved organic material from water by filtration of these materials that are physically too large to pass through the pores of the membrane.

Feed - Raw or pretreated water under pressure entering a membrane system.
Flux - Flow or flow rate of permeate through a membrane or membrane element which is the amount of water produced per day or per square foot of membrane surface area.
Permeate - Portion of the feed stream which passes through the membrane and is consumed.
Element - A usable configuration into which the membrane material is formed.
Module - Combination of membrane elements and the housing or pressure vessel into which the membrane is inserted for use in an operating system.
Concentrate, reject, or brine - Contaminants or impurities rejected by a membrane and carried away by a water stream.
Product Recovery - Percentage of the feed water flow that passes through the membrane as the permeate.

Reverse osmosis module designs consist of hollow fiber, spiral wound, tubular, and plate and frame.[20,21,24] The most common modules for drinking water treatment are the hollow fiber and spiral wound, which are described below.

Hollow Fiber

Semipermeable hollow fiber membranes are usually produced using aromatic polyamides (Dupont) or cellulose triacetate (Dow Chemical). The membrane material is spun into hairlike hollow fibers having an outer diameter of 85 to 200 um. The fibers are bundled together in either a U-shaped configuration for brine flow on the outside or in a straight configuration for brine flow in the inside. The fibers are wrapped around a support frame, and the open ends of the looped fibers are epoxied into a tube sheet.[25] Figure 6 shows a typical hollow fiber module.

For most operations, raw water is pumped under pressure (200-400 psig) through a distributor tube and flows outward through the fiber bundle. A portion of the pressurized feed water permeates through the wall of each hollow fiber and into the bore, leaving most of the dissolved solids, organics, and bacteria in the concentrated reject water. The permeate forced into the fiber bores is withdrawn at the epoxy tube sheet end of the membrane shell. A flow screen inserted between the bundle and shell permits the concentrate to exit the shell through a reject port.

Pressure vessels normally contain a single element. The elements have high packing density of 39,400 m^2/m^3 (12,000 ft^2/ft^3) and resistance to pressure collapse even at 600-1000 psig.[25] Broken fibers are claimed to be self healing by collapsing.[25] Product recovery is 50-60% of the feed flow rate.[25,26] The manufacturing and equipment cost is relatively low.

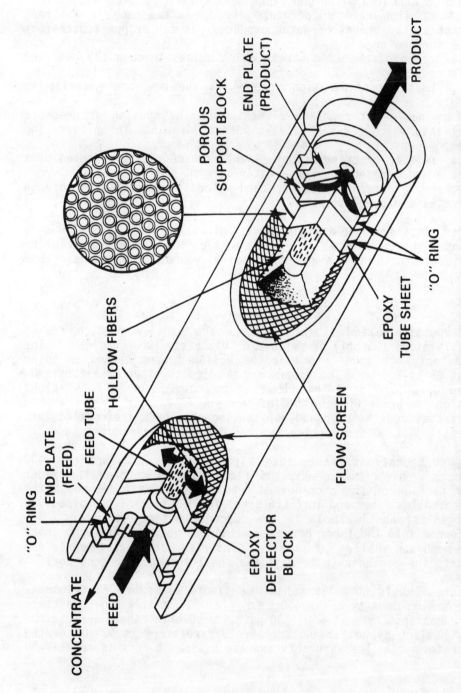

FIGURE 6.   HOLLOW FIBER MEMBRANE MODULE

Spiral Wound

The spiral wound module contains two layers of semipermeable membranes separated by a woven fabric. Figure 7 shows a typical spiral wound module. The semipermeable membranes can be constructed of various materials including cellulose acetate, cellulose triacetate, or thin-film composites. The woven fabric can consist of nylon or Dacron or, in the case of the thin-film composites, a polyester web with a polysulfone coating.

A flexible envelope is formed by sealing the edges of the membrane on three sides, with the fourth open side attached to a perforated tube. A sheet of plastic netting placed adjacent to the membrane envelope separates the membrane layers and promotes turbulence in the feed stream during operation. The envelope and netting are wrapped around the central tube in a spiral configuration. Pressurized feedwater permeates through the membrane into the fabric, where it is directed to the perforated central tube for collection and removal as product water while the concentrate stream passes out the other end of the element. The elements are housed in a cylindrical vessel with the feed flowing in a straight-axial path parallel to the direction of the permeate collection tube.

Spiral wound modules are generally not subject to sealing and fouling, but highly turbid waters will require pretreatment. Product recovery per element ranges from 5% to 15% of the feed flow rate. The spiral wound pressure vessel can house up to six membrane elements in series and the spiral wound configuration offers a good ratio of surface area to volume.

Membrane processes for potable water treatment are divided into two basic categories.[27] One category is that in which the membrane is water permeable. This category includes processes such as reverse osmosis, nanofiltration, ultrafiltration, and microfiltration. The particle size that can be rejected by these typical separation processes is shown in Figure 8. The other category where the membrane is impervious to water includes processes such as electrodialysis and electrodialysis reversal.[27] The two membrane systems generally used in water treatment are reverse osmosis and ultrafiltration.

Reverse Osmosis

Reverse osmosis is not a process that removes chemicals from water, as in adsorption but instead it rejects compounds based on their molecular properties and characteristics of the reverse osmosis membrane. Contaminant removal in a reverse osmosis process is obtained by passing contaminated water through a semipermeable membrane. The term semipermeable is used because it allows the passage of water molecules while blocking most dissolved and suspended molecules.[1] The flow of water through a semipermeable membrane is in the direction of decreasing water potential.

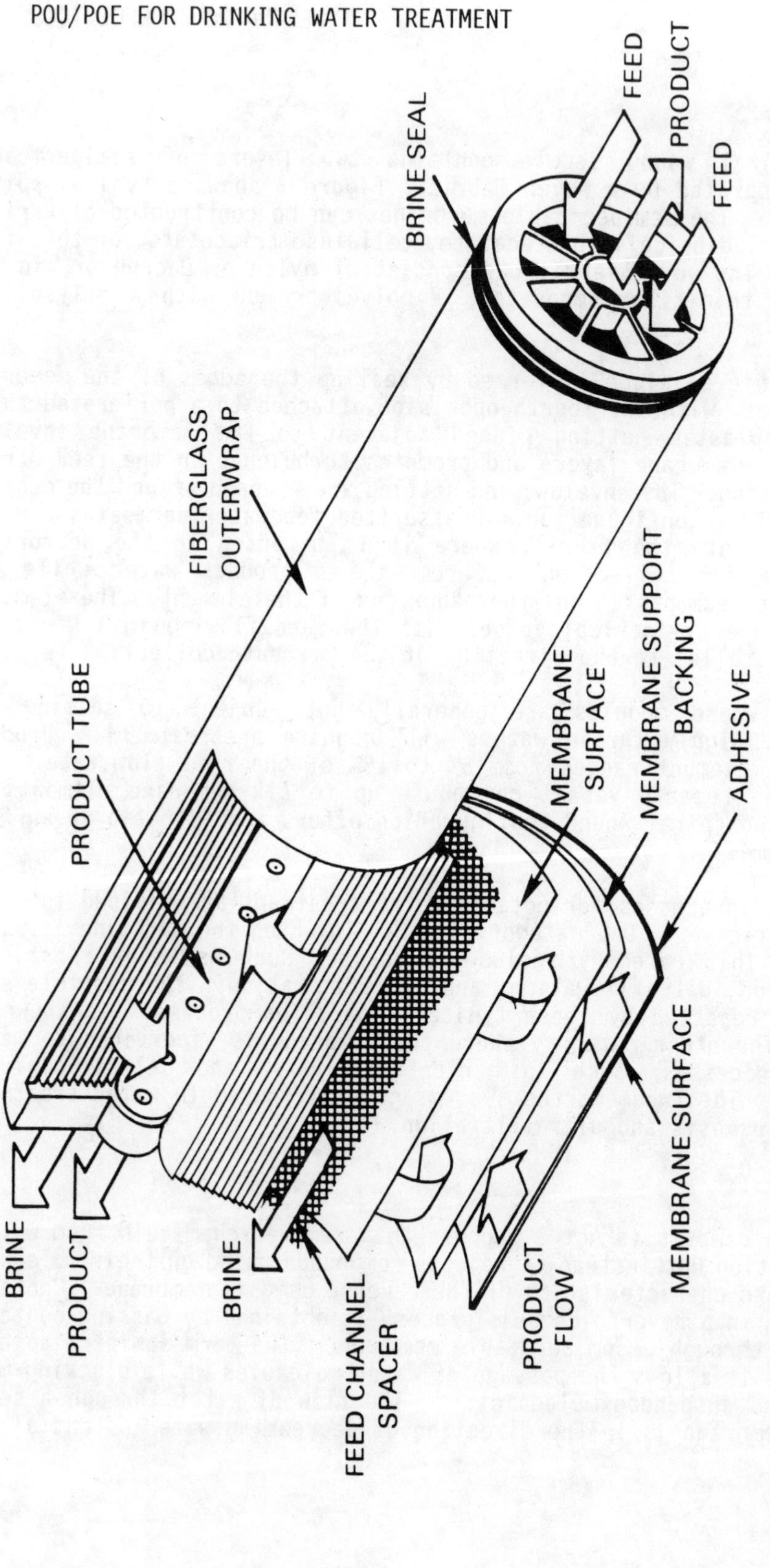

FIGURE 7.   TYPICAL SPIRAL WOUND MODULE

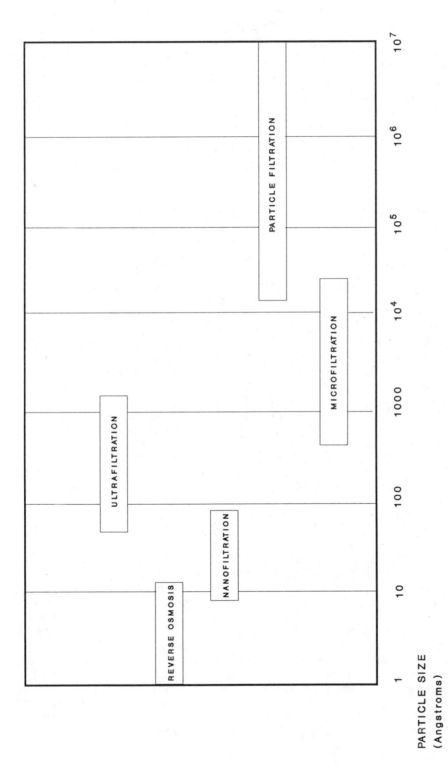

Figure 8.   Typical Separation Processes

With contaminated water containing dissolved ions on one side of the semipermeable membrane and essentially clean water on the other side of the membrane, the driving force for the two flows is the difference in chemical potential between the two solutions.[28]   Increasing the dissolved ion concentration of a water will decrease its osmotic potential, and hence decreases its water potential.[1]   In the reverse osmosis process, water on the feed side of the membrane has a lower water potential than the product side because of its high contaminant concentration.   The natural direction of flow (osmosis) is therefore, from the product side of the membrane to the feed side.   This flow is reversed by applying enough pressure to the feed water to overcome its loss in osmotic potential. This reversal of natural flow has given the process the term reverse osmosis.   Figure 9 shows how reverse osmosis works.[20,28,29,30]

Reverse osmosis has been successful in rejecting inorganic ions, turbidity, bacteria, and viruses.[24]   Some work has been performed on organic compounds with varying degrees of success.[22,31]   The literature indicates success in rejecting chemicals with molecular weights greater than 200.[24]   In general, reverse osmosis membranes will reject 90-98% of dissolved salts and 98-100% of dissolved organic contaminants with molecular weights above 200 daltons.[20]   Other organic compounds are not as well removed and some are not removed at all. Removal efficiencies are usually dictated by the type of membrane material used as discussed previously. Also, most point-of-use/ point-of-entry units contain either an activated carbon filter prior to or after reverse osmosis treatment making it difficult to determine the effectiveness of reverse osmosis alone.   Some factors that can affect membrane performance include raw water quality parameters such as pH, temperature, and turbidity.   For example, cellulose acetate membranes have a pH tolerance of 2 to 8 and require disinfection because of poor resistance to bacteria. Conversely, thin-film composite membranes have a pH tolerance of 2 to 12 and are resistant to bacteria. Most reverse osmosis units operate with a 10-30% recovery rate by can have a staged multiple units to obtain higher recoveries.   A typical point-of-use reverse osmosis unit is shown in Figure 10.

Ultrafiltration

Ultrafiltration membranes are specially formulated and constructed to remove particulate and organic impurities by the sieving process using an array of pores in the membrane surface.[23]   While these membranes can separate virtually all particulate matter, microorganisms, and larger organic molecules from water, they have little effect on separating out dissolved solids such as mineral salts. The mechanism of ultrafiltration is closely associated with typical filtration in that the dissolved organic materials are removed because they are physically too large to pass through the pores of the membrane.[32]   However, there are considerable differences between ordinary filtration and ultrafiltration in that a velocity vector must exist parallel to the plane of ultrafiltration.[33] This is generally provided by a flow of fluid across the membrane with the separation of the fluid into a permeate and concentrate.

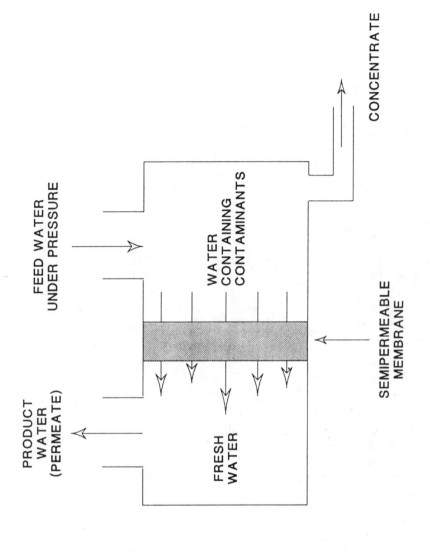

Figure 9. How Reverse Osmosis Works

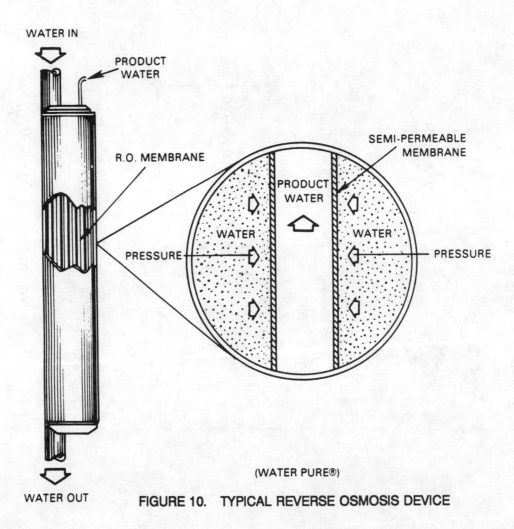

(WATER PURE®)

FIGURE 10.   TYPICAL REVERSE OSMOSIS DEVICE

Ultrafiltration has been used in industry to exclude large organic molecules either for purification of the permeate or concentration of a marketable retenate. Until recently ultrafiltration has received little attention as a drinking water treatment process. Ultrafiltration membrane materials are sometimes the same as those used for reverse osmosis but they receive a modified treatment to give them a different structure.[33] Available membranes include such polymers as cellulosic, polysulfone, thin film composites, and polyvinylidene fluoride.[34] Spiral wound, tubular, and hollow fiber ultrafiltration systems have been developed.

Ultrafiltration membranes typically have pore sizes ranging from 40 to 1,000 Angstroms and may be employed for the removal of submicron colloidal particles, microorganisms, silt and large molecular weight organic compounds. All reverse osmosis membranes have ultrafiltration capabilities and some of the first ultrafiltration membranes used in drinking water treatment applications were reverse osmosis membranes that were operated at lower feed pressures (<200 psig). Ultrafiltration is usually applied to the membrane separation of solutes in the molecular weight range of 100 to 1,000. Although individual ultrafiltration membranes may only produce product recoveries around 20 to 30%, depending on operational conditions such as pressure and flux, the membrane system can be staged to produce much higher recoveries that can exceed 80%. For a point-of-entry system one might possibly use the configuration shown in Figure 11 where the concentrate from the first stage of membranes is subsequently passed into two other membrane modules.

Ion Exchange

Ion exchange works by exchanging charged ions in the water for ions of similar charge on the resin surface. In cation exchange, the ions most often displaced from the resin are sodium ions. For anion exchange, the ion exchanged is usually chloride. Most of the inorganic compounds can be removed by ion exchange but most of the organic compounds commonly found in sources of drinking water cannot be effectively removed. Ion exchange has been used to remove compounds that cause aesthetic rather than health related problems. Ion exchange systems are commonly called water softeners.

Cation softeners substitute positively charged sodium ions attached to the resin surface for positively charged ions such as calcium and magnesium (which cause water to be classified as hard), iron, manganese, other positively charged metals, and 90 to 97 percent of cationic radioactivity.[1] The cation softeners can eliminate such things as soap scum that often appears on bathtubs, reduce soap film on skin, hair, and dishes, eliminate stains on plumbing, and deter scale buildup in pipes. Shown below is what happens in a cation softener.

$$CaX_2 + M_gY_2 + Resin(Na^+) \rightarrow NaX + NaY + Resin (Ca^{++} \text{ and } Mg^{++})$$

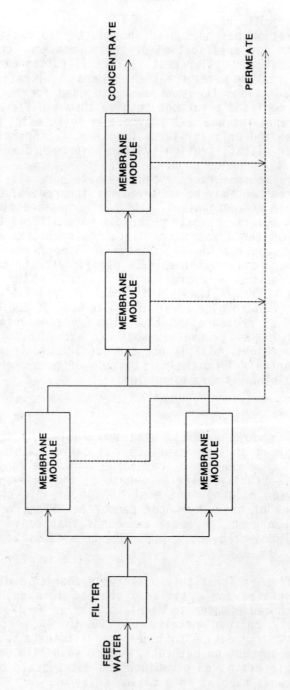

Figure 11. Possible Configuration of Membrane System to Increase
Product Recovery

A basic softener design is shown in Figure 12, while Figure 13 shows what happens within the resin tank.

After a period of use, the resin will become saturated with contaminant ions and regeneration is required. Regeneration is accomplished by introducing a highly concentrated sodium chloride solution into the unit to reverse the exchange. The contaminant ions, such as calcium and magnesium, which had attached to the resin are removed by the concentrated sodium chloride solution and replaced by the original resin ion (sodium). The displaced contaminant ions are disposed to the drain with the regenerant brine and the unit is ready for operation again.

The most common applications for anion exchange systems are for the removal of nitrates and dealkalization (such as bicarbonate removal). Other anions are also removed such as arsenic, hexavalent chromium, selenium, and sulfate. The anion softener design is similar to the cation softener. The only difference is the resin.

Distillation

Distillation is a process of driving off gas or vapor from liquids by heating and then condensing the gas or vapor as shown in Figure 14. Therefore, distillation is a process that uses evaporation to purify water. Water that contains contaminates is heated to form steam. Dissolved minerals (inorganic compounds), nonvolatile organics, and particulates do not evaporate with the water and are left in the boiling tanks. The steam is then cooled and condenses to form purified water. There are two types of home water distillation systems: air-cooled and water-cooled. Air-cooled distillers are reported to make one gallon of distilled water from one gallon of tap water whereas water cooled are reported to make one gallon of distilled water from 5 to 15 gallons of tap water depending on the system design.[35] Others have reported production rates of 3 to 11 gallons per day.[36] Examples of two types of distillation units are shown in Figures 15 and 16.

Distillation is most effective in removing inorganic compounds such as metals (iron and lead) and nitrates, hardness which is mainly calcium and magnesium, and particulates from contaminated water.[36] Most bacteria and some viruses in the feed water can be killed by the high temperatures. Those not killed are separated from the water as the steam rises from the tank. The effectiveness of distillation in removing organic compounds varies depending on the chemical characteristics of the compounds such as water solubility and boiling point. It has been reported that organic compounds that boil at temperatures greater than the boiling point of water such as some pesticides can be effectively removed from water by distillation.[36] Also, volatile organic chemicals will evaporate at the high temperatures used in distillers and have been reported to be removed by venting them to the atmosphere.[1] However, if these organics are not removed prior to condensation, they will be carried over into the condensate and recontaminate the purified water. Any claims for removal of organics such as chloroform, trihalomethanes, pesticides, herbicides, etc.

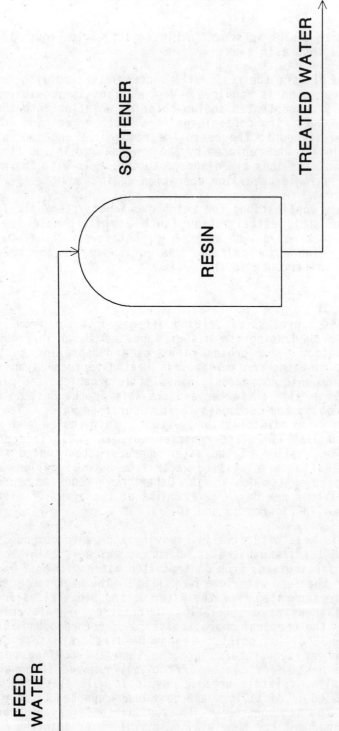

Figure 12. Typical Ion Exchange System

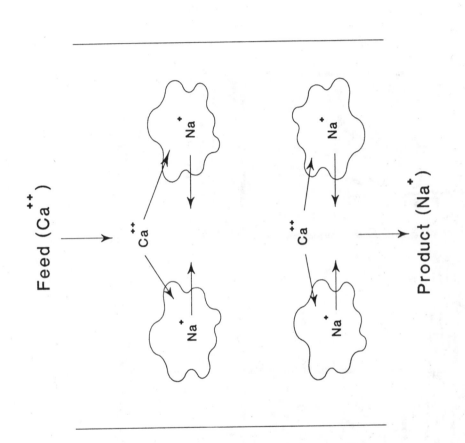

Figure 13. The Ion Exchange Process

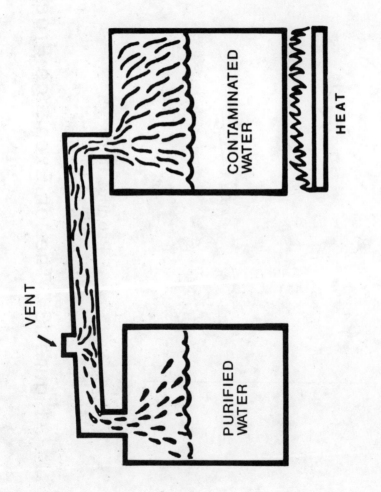

FIGURE 14.   THE DISTILLATION PROCESS

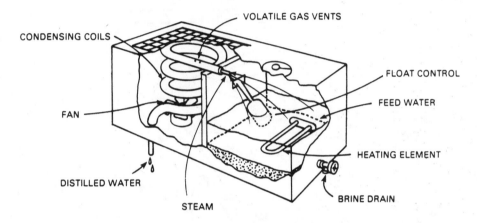

**FIGURE 15.   EXAMPLE OF AN AIR COOLED
DISTILLATION UNIT**

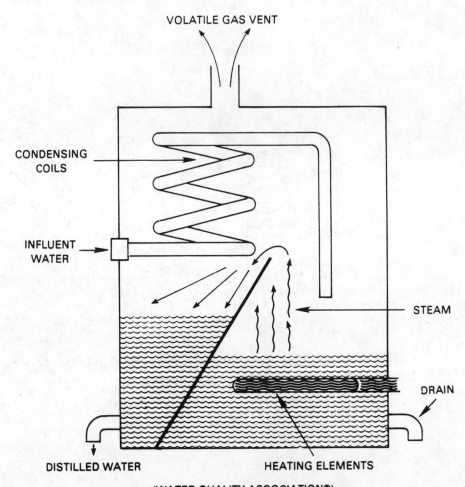

VOLATILE GAS VENT

CONDENSING
COILS

INFLUENT
WATER

STEAM

DRAIN

DISTILLED WATER   HEATING ELEMENTS

(WATER QUALITY ASSOCIATION®)

**FIGURE 16.   EXAMPLE OPERATION OF A
DISTILLATION UNIT**

should be backed by test data.[37] Distilling units have relatively high electrical demands which require about 3 kilowatt-hours per gallon of water treated.[1] Some units contain preheat tanks and automatic shutoffs to minimize these power requirements.

The beneficial effects of distillation may be debateable. For instance, one may argue that the removal of metals, bacteria, and other contaminants far out-weighs the benefits of ingesting some minerals that are reportedly beneficial. By removing minerals such as calcium and magnesium the water becomes flat-tasting. This water is usually referred to as soft water. Soft water has been implicated as a major contributor to an increase in cardiovascular disease when compared to hard-water (high mineral content consisting of calcium and magnesium).

Air Stripping

Packed tower aeration is a proven technology for removing volatile organic chemicals from drinking water supplies.[24] In air stripping, untreated water is usually introduced at the top of a column. This water flows down by gravity while air is pumped upward by a mechanical blower. The column contains packing material which increases the area of the air-liquid interphase for better mass transfer. A typical configuration of packed tower air stripping system is shown in Figure 17. Volatile organic chemicals are transferred from the water to the air, which is then vented outside. Volatile organics removal from water by air stripping depends on several factors: (1) hydraulic loading rate, (2) air-to-water ratio, (3) type of packing material, (4) height of packing materials, and (5) type and concentration of the volatile organics to be removed. Storage and repumping facilities are needed after air stripping to distribute the treated water.

Air stripping is being used for point-of-entry applications, especially where high concentrations of volatile organics have to be removed from drinking water. Many point-of-entry systems have been installed to remove volatile organics from contaminated groundwater. Oftentimes this contamination has been caused by spills of industrial solvents, leaking underground gasoline or oil storage tanks, and hazardous waste or landfill sites. By using granular activated carbon alone, the volatile organics can be removed, but only for short periods of operation before the carbon has to be replaced. For this reason, air stripping has become a viable treatment option, either by itself or in combination with granular activated carbon. A diagram of a home air stripper and GAC filters in operation in a home in Elkhart, Indiana is shown in Figure 18.

Another use for air stripping has been the removal of methane gas. This gas has been found in wells where landfills have existed or oil and gas drilling is prevalent. Methane gas present in drinking water is not widespread nor is it a health problem. However, explosions can occur if it is in high enough concentrations and it can give water a milky look.[38]

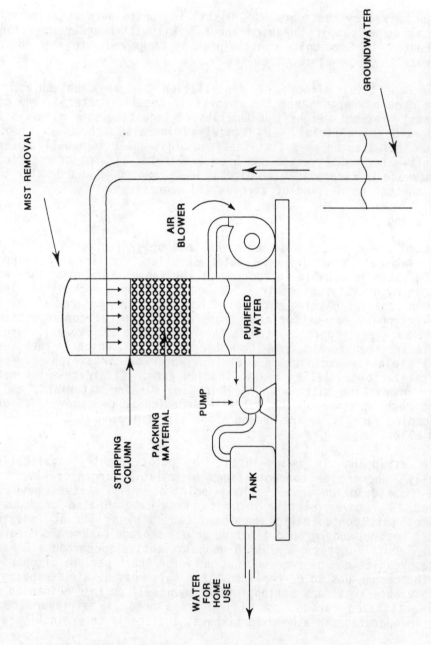

Figure 17. Typical Configuration of Home Air Stripping System

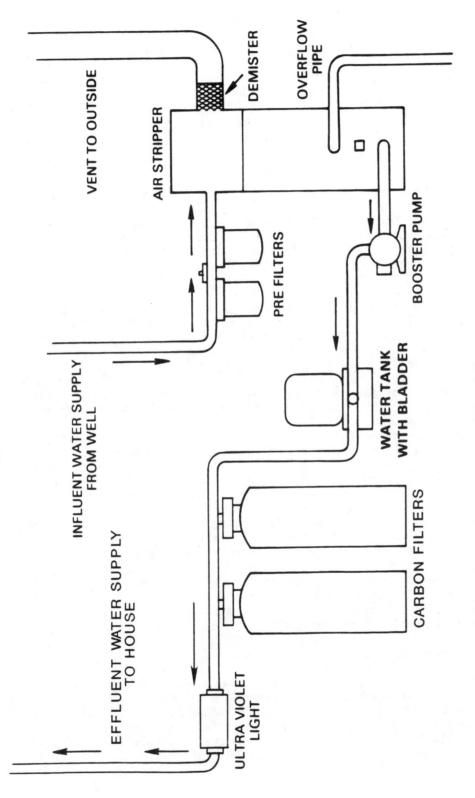

FIGURE 18.   HOME AIR STRIPPER AND GAC FILTER

Radon gas can also be removed by aeration. Packed tower aeration removals are limited to the home ceiling height if placed inside. At a packing height of six feet, removals of 90 to 95 percent can be obtained. Higher removals up to 99 percent can be achieved if greater packing height is used. Because radon is removed from the water and placed into the air, caution is required in placement of the discharge air vent. Also, other aeration methods such as diffused bubble aeration produced removals greater than 99 percent.[19]

Activated Alumina

Activated alumina is most often used to remove fluoride from drinking water although it can also remove arsenic, chromium, selenium, and inorganic mercury.[1] The fluoride removal mechanism by activated alumina is an exchange/adsorption process. The surface characteristics of activated alumina are important in producing the adsorptive properties of the alumina. The granular and porous nature of activated alumina are shown in Figures 19 and 20.

Activated alumina is hydrated alumina (Al_2O_3) which has been heat treated to approximately 750°F. The electrostatic attraction between the contaminant molecules in the influent or feed water and the adsorption media is the driving force of the exchange/adsorption process. One of the advantages of activated alumina treatment for fluoride removal is that the alumina is highly preferential to fluoride and the medium's capacity is not normally reduced significantly by competition from sulfate, chloride, or bicarbonate.[1] The optimum pH range for fluoride removal with activated alumina is 5 to 6 for most waters. Operation at pH values either higher or lower than this range can substantially decrease the exchange capacity of the activated alumina.

The transfer of fluoride from a drinking water source to the surface of activated alumina has been described as a four step process: (1) mass transfer (diffusion) from the drinking water to the alumina, (2) film diffusion at the outer surface of the alumina, (3) diffusion through the alumina pores, and (4) sorption onto the surface.[39] Film diffusion and diffusion through the water can be controlled somewhat by varying the hydraulic loading rates or flow. Therefore, the rate limiting factor is pore diffusion after the outer surface of the activated alumina is initially covered. After the activated alumina bed reaches its adsorption capacity, the spent cartridge will have to be replaced and should be disposed of properly or regenerated.

Disinfection and Oxidation

Although it is very important to remove organic and inorganic contaminants from drinking water to control or eliminate chronic health effects that are caused by ingesting these contaminants, it is oftentimes equally and sometimes more important to control or eliminate microbiological contaminants that can cause acute health problems. This means using a disinfectant for bacteria, virus, and cyst control when these are

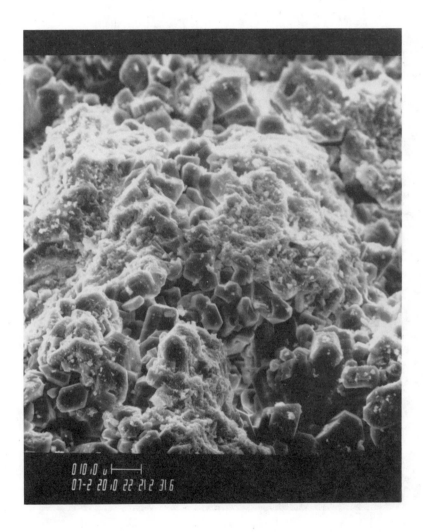

Figure 19.   Activated Alumina Granules (10 microns-1400X Magnification)

Figure 20. Activated Alumina Granules
(1 Micron - 6,600X Magnification)

present in the source water.   Basically, all of the treatment processes discussed previously should be followed by adequate disinfection to ensure that the consumer receives a microbiologically safe water.

Disinfectants that are usually used in point-of-use/point-of-entry systems are ultraviolet light, ozone, chlorine, silver impregnated carbon, and iodine.   Ultraviolet light is a popular disinfection method in combination with other treatment techniques.   One of the major advantages of ultraviolet disinfection is that it disinfects without the addition of chemicals.   Therefore, no taste, odor, or chemical by-products are produced.   Ultraviolet light also works rapidly and the equipment requires little maintenance.   Changes in flow rates within the operating range of the ultraviolet unit present no problem.   An overdose of ultraviolet light presents no danger and is in fact considered a safety factor.[40]

Ultraviolet Light.   Ultraviolet light is only one region of the electro-magnetic spectrum as shown in Figure 21.[41]   Wavelengths used for killing microbes are between wavelengths of 200 and 300 nm.   However, wavelengths of 260 to 265 nm are most effective in killing microbes.   Low pressure mercury lamps with a quartz bulb are the most common sources used to produce ultraviolet radiation within the effective germicidal wavelength. These lamps are similar in design, construction, and operation to ordinary fluorescent lamps.   The major difference is that the ultraviolet lamps are constructed of ultraviolet transmitting quartz whereas the fluorescent lamp has soft glass with an inside coating of phosphorus which converts ultraviolet to visible light.   The quartz tube transmits 93% of the lamp's ultraviolet energy while the soft glass emits very little.   A typical ultraviolet lamp is shown in Figure 22.[41]   Over 85% of the ultraviolet radiation emitted from these lamps is at 253.7 nm, close to the optimum germicidal wavelength.[40]

Manufacturers produce ultraviolet systems ranging in size from 0.5 gpm to several hundred gpm.   Most ultraviolet manufacturers claim that their systems have an effective life of between 6,000 and 12,000 hours. The germicidal effect of an ultraviolet lamp is the result of ultraviolet light penetration through bacteria cell walls and adsorption by the nucleic acids which cause genetic damage.

In an ultraviolet system, water enters the purifier and flows into the space between a quartz sleeve which contains the ultraviolet lamp and the outside chamber wall.   The quartz sleeve is used to separate the lamp from the water to maintain lamp operating temperature.   Wiper segments are used to induce turbulence in the flowing liquid to ensure uniform exposure of microorganisms to the ultraviolet light.   A wiper assembly facilitates periodic cleansing of the quartz sleeve without any disassembly or interruption of operation.   With a sight port, visual observation of lamp operations is possible.   A typical ultraviolet purifier is shown in Figure 23.[42]

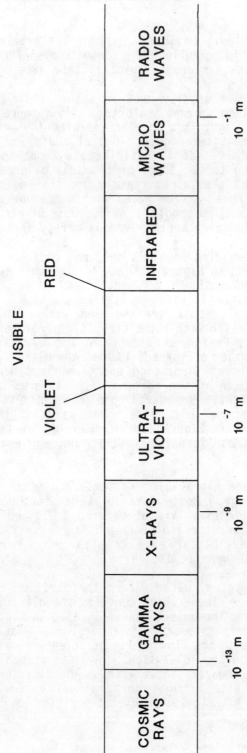

Figure 21. Electromagnetic Spectrum

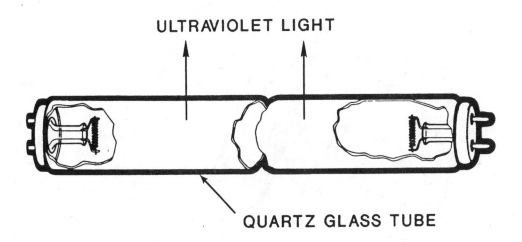

FIGURE 22. TYPICAL ULTRAVIOLET LAMP

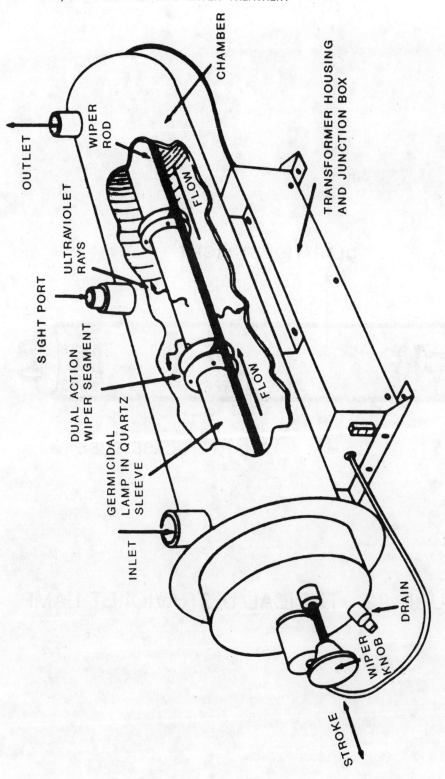

FIGURE 23.   TYPICAL ULTRAVIOLET PURIFIER

Ozone. The primary uses of ozone in drinking water treatment have been for disinfection and oxidation of inorganics and organics. Ozone is a highly reactive and unstable gas. Because of its instability, ozone has to be generated and used on-site as needed. The half-life of ozone in water varies with the water quality but is generally no more than a few hours and may typically be a few minutes. At this time, ozone is generated commercially by two procedures; ultraviolet radiation and corona discharge.[43]

Some older ozone generators for point-of-use/point-of-entry applications use ultraviolet radiation because they are the lowest cost on a per unit basis. For point-of-use/point-of-entry applications air is used as the generator feed gas which means that only a portion of the oxygen in the air is converted to ozone. For the ultraviolet radiation method, ozone concentrations of 0.1% to 0.01% by weight are produced. The corona discharge method produces ozone concentrations of 1% to 3% by weight. The ultraviolet radiation generators are less expensive because they do not dry the ambient air whereas with corona discharge there is drying of the ambient air. The corona discharge generators are more costly on a per unit basis but they produce more ozone at constant output rates and at higher ozone output concentrations.[43]

Also of importance is the rate of ozone production by the two types of generators. Currently available ultraviolet tubes of 185 nm radiation produce a maximum of 0.5 g/h of ozone per tube. The 254 nm ultraviolet tubes produce less ozone (about 0.3 g/h).[43] Corona discharge ozone generator tubes fed with properly dried air, however, produce around 2 g/h of ozone and by arranging tubes together in modules, generation rates up to 36 g/h can be achieved by a single ozone generator.[43]

The ozone production rate and the concentration of ozone produced depends on the dryness of the air feed gas. For corona discharge generators, suppliers recommend that feed gas air be dried to at least minus 60°C dew point. This dew point maximizes ozone production. When ozone is generated by corona discharge there are five important components to the ozonation system: power supply, air preparation, ozone generator, ozone contactor, and off-gas destruction.[43]

For many point-of-use/point-of-entry applications, two stages of ozonation may be appropriate. These two stages are used for oxidation of contaminants and for disinfection prior to use of the water. A possible ozone treatment system is shown in Figure 24.

Chlorine. The most widely used water disinfectant in the United States is chlorine. When placed into water, chlorine forms hypochlorous acid ($HOCl$) and hypochlorite ion (OCl^-) with the amount of these determined by the pH of the solution. It is important to maintain as much $HOCl$ as possible because it is about 100 times more powerful of an oxidant and disinfectant than OCl^-. Besides pH, chlorine's disinfection and oxidation capabilities are also influenced by organic and inorganic compounds which can cause a chlorine demand.

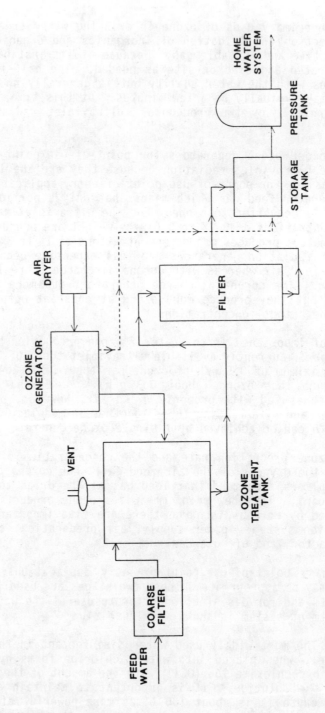

Figure 24. Ozone Home Water Treatment System for Oxidation
And Disinfection

Public water systems use chlorine in the gaseous form but this is considered too dangerous and expensive for home use. Private systems use liquid chlorine (sodium hypochlorite) or dry chlorine (calcium hypochlorite) To avoid hardness deposits on equipment, manufacturers recommend using soft, distilled, or demineralized water when making chlorine solutions.[44] The most common form of liquid chlorine comes from household bleach which contains about 5.25% available chlorine. However, there are concerns about contaminants in the bleach such as heavy metals. Up to 18% available chlorine can be obtained from commercial laundry bleach. The dry chlorine is a powder which is dissolved in water and produces about 4% of available chlorine.

Equipment for continuous chlorination of a private water supply has been described by Wagenet and Lemley.[44] A description of a pump type and injector chlorinator follows. The pump type chlorinator, which is commonly used, is a positive displacement device. This pump adds small amounts of chlorine to the water with the dose either fixed or varied with water flow rates. The chlorine is drawn into the device and then pumped to the water delivery line. This type of chlorinator is recommended for low and fluctuating water pressure. A diagram of a pump type chlorinator is shown in Figure 25.

The injector or aspirator chlorinator is a simple, inexpensive mechanism. It requires no electricity but instead a vacuum is created by water flowing though a tube which draws chlorine into a tank where it mixes with untreated water. One of the problems with this type of chlorinator is that the chlorine dose is not always accurate. Once treated the solution is fed into the water system. A diagram of an injector or aspirator type of chlorinator is shown in Figure 26.

Iodine. There are various methods of application of iodine for treatment of water. Disinfection appears to be the main reason for using iodine since it will not precipitate iron or manganese and its oxidizing power for taste and odor removal is more limited than that of chlorine. However, the disinfecting ability of iodine is not affected as much as chlorine by high pH or the presence of organic or other nitrogen-containing substances.[37] Elemental iodine and hypoiodous acid are effective disinfecting agents. Although chlorine is a more potent microbiological control disinfectant than iodine, it is often more difficult to maintain a chlorine residual, especially in the presence of ammonia. Therefore, since iodine does not combine with ammonia, a free iodine residual is established easier and maintained longer. There appears to be two types of iodine units that are presently used in point-of-use/point-of-entry systems.[45]

One type of system uses an iodine dosing method which delivers a stream of saturated iodine solution into the water supply stream at a specified dosage. The iodine dosing method consists of a rather simple

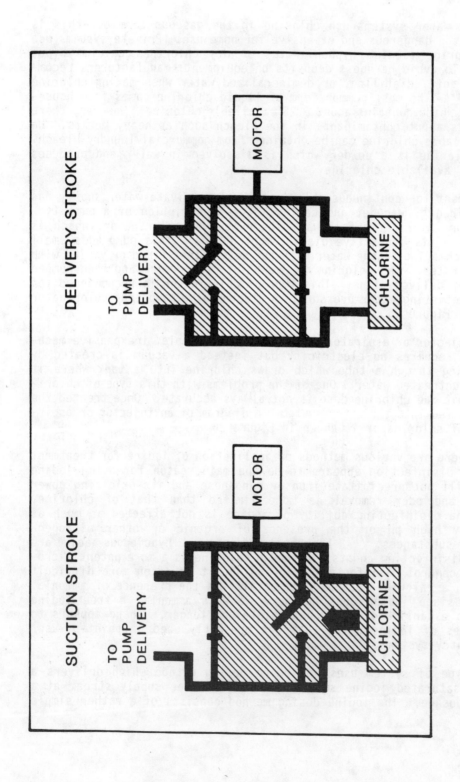

FIGURE 25. PUMP TYPE CHLORINATOR

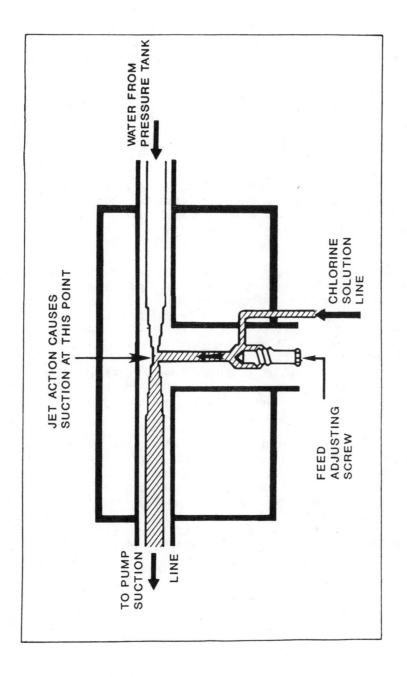

WATER FROM PRESSURE TANK

JET ACTION CAUSES SUCTION AT THIS POINT

TO PUMP SUCTION LINE

CHLORINE SOLUTION LINE

FEED ADJUSTING SCREW

FIGURE 26. INJECTOR CHLORINATOR

device consisting of passing some water through a bed of elemental iodine crystals. Detention of this water in the iodine bed is maintained long enough to reach saturation. This iodine solution is then injected into the water supply system.

Using a valving system, a small portion of the water supply flow passes through a tank of iodine crystals forming a saturated iodine solution. The saturated solution is introduced into the water supply where it is diluted to a desired concentration of 0.5 to 1.0 mg/L. At normal operating temperatures of 5°C to 20°C, a detention time of 15 minutes is recommended requiring a storage tank of sufficient volume to achieve this contact time. However, excessive detention times (in excess of 6 hours) will tend to cause the free iodine residual to dissipate.[37] Residual iodine concentrations above 1.5 mg/L should be avoided so that taste and odors and potential thyroid problems will not occur. A diagram of an iodine-dosing unit is shown in Figure 27.[45]

The other type of iodine system is an iodine resin. These are iodinated strong-base resins that release iodine upon demand to disinfect. The original demand-release iodinated resins were called Triocide.[46] A resin triiodide unit was tested by the Canadian Department of National Health and Welfare.[45] This unit consisted of a woven fiber filter to remove small particles to 10 μm. A second cartridge contained granular activated carbon to remove taste and odor. The third cartridge contained the triiodide resin in a sealed disposable unit containing a polystyrene divinylbenzene quaternary ammonium anion exchange resin that had been substituted with the triiodide ion. Figure 28 shows the triiodide resin cartridge components.

A new technique has been developed where the triiodide resin is further saturated with elemental iodine until the triiodide is converted to pentaiodide. This resin called Pentacide is reported to be 1,000 times more potent than Triocide which is more potent than iodine.[46] There is a concern with the use of both of these units because of the residual iodine in the water which may be a problem for people with thyroid conditions. For them to be acceptable as continuous disinfectant systems, methods need to be developed to remove residual iodine from the water.

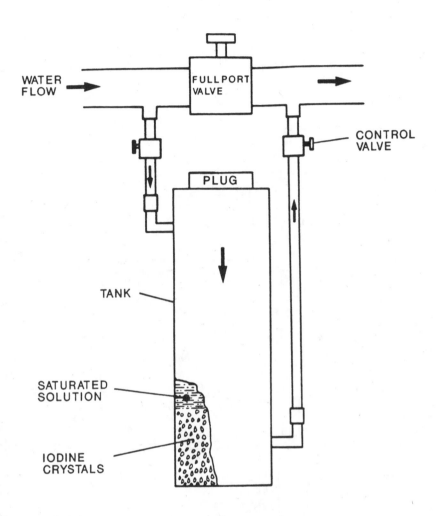

FIGURE 27. IODINE - DOSING UNIT

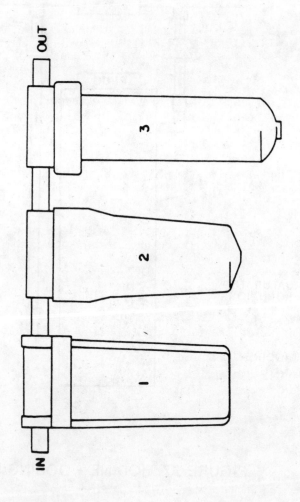

FIGURE 28. RESIN TRIIODIDE SYSTEM COMPONENTS
(1. FINES FILTER, 2. ACTIVATED CARBON
FILTER, 3. RESIN TRIIODIDE CARTRIDGE)

REFERENCES

1. Guide to Point-of-Use Treatment Devices for Removal of Inorganic/ Organic Contaminants from Drinking Water, New Jersey Department of Environmental Protection, (1985).

2. Lykins, Jr., B. W., Clark, R. M., and Westrick, J. J., "Treatment Technologies for Meeting U.S. Drinking Water Regulations", Paper presented at 1990 Joint Annual Conference of Ontario Section AWWA/ Ontario Municipal Water Association, Toronto, Ontario, Canada, May 6-9, 1990.

3. Sorg, T. J., "Methods of Removing Drinking Water Contaminants and Their Limitations: Inorganics and Radionuclides", Paper presented at the Water Quality Association Annual Convention, San Antonio, Texas, March 14-18, 1990.

4. Bellen, G., Anderson, M., and Gottler, R., "Management of Point-of-Use Drinking Water Treatment Systems", U.S. Environmental Protection Agency Contract No. R809248010, (June 1985).

5. Cole, R. W., "Effectiveness of Commercially Available Home Water Purification Systems for Removing Organic Contaminants", Air Force Engineering and Services Center, Tyndall Air Force Base, FL, (1986).

6. "Process Design Manual for Carbon Adsorption", EPA 625/1-71-002a, (October 1973).

7. Sontheimer, H., Crittenden, J. C., and Summers, R. S., "Activated Carbon for Water Treatment", AWWA Research Foundation, (1988).

8. Culp, R. L., Wesner, G., and Culp, G. L., "Handbook of Advanced Wastewater Treatment", Van Nostrand Reinhold Co., New York, (1978).

9. Dyke, K. V. and Kuennen, R. W., "Performance and Applications of Granular Activated Carbon Point-of-Use Systems", Proceeding: Conference on Point-of-Use Treatment of Drinking Water, Cincinnati, OH, October 6-8, 1987.

10. Greenbank, M. and Manes, M., "Application of the Polany: Adsorption Potential Theory to Adsorption from Solutions on to Activated Carbon. 12. Adsorption of Organic Liquids from Binary Liquid-Solid Mixtures in Water", *Journal Physical Chemistry*, 86(21):4216-4223, 1982.

11. McGuire, M. J. and Suffet, I. H., "The Calculated Net Adsorption Energy Concept", Activated Carbon Adsorption of Organics from the Aqueous Phase, Vol. 1, M. J. McGuire and I.H. Suffet, Eds. (Ann Arbor, MI: Ann Arbor Science Publishers, Inc.), 91, (1980).

12. Fredenslund, A., Jones, R., and Prausnitz, J., "Group-Contribution Estimation of Activity Coefficients in Nonideal Liquid Mixtures", *Journal American Institute Chemical Engineering*, 21(6):1086-1099, (1975).

13. Speth, T. F., "Removal of Volatile Organic Compounds from Drinking Water by Adsorption", Significance and Treatment of Volatile Organic Compounds in Water Supplies (N. M. Ram, R. F. Christman, and K. P. Cantor, Eds.), Lewis Publishers, Chelsea, MI, (1990).

14. Kong, E. J. and DiGiano, F. A., "Competitive Adsorption Among VOCs on Activated Carbon and Carbonaceous Resins", *Journal American Water Works Association*, 78(4):181-188, (1986).

15. Jain, J. S. and Snoeyink, V. L., "Competitive Adsorption from Biosolute Systems on Activated Carbon", *Journal Water Pollution Control Federation*, 45:2463-2479, (1973).

16. Randke, C. J. and Prausnitz, J. M., "Thermodynamics of Multi-Solute Adsorption from Dilute Liquid Solutions", *Journal American Institute Chemical Engineering*, 18:761-768, (1972).

17. Yen, C. and Singer, P. C., "Competitive Adsorption of Phenols on Activated Carbon", *Journal Environmental Engineering Division*, ASCE, 110(5):976-989, (1984).

18. Critten, J. C., Luft, P. Hand, D., Oravitz, J., Loper, S., and Ari, M., "Prediction of Multicomponent Adsorption Equilibria Using Ideal Adsorbed Solution Theory", *Environmental Science and Technology*, 19(11):1037-1043, (1985).

19. Lowry, J. D., Brutsaert, W. F., McEnerney, T., and Molk, C., "Point-of-Entry Removal of Radon From Drinking Water", *Journal American Water Works Association*, (April 1987).

20. Cartwright, P. S., "Understanding Reverse Osmosis", *Water Conditioning and Purification*, 29(7), (August 1987).

21. Slovak, J. and Slovak, R., "A Guide to Reverse Osmosis Membranes and Modules", *Water Conditioning and Purification*, 25(8):17, 26-27, (September 1983).

22. Fronk, C. A., Lykins, Jr., B. W., and Carswell, J. K., "Membranes for Removing Organics from Drinking Water", submitted to American Filtration Society/Fluid Particle Separation Journal, December 1990.

23. Slovak, J. and Slovak, R., "Developments in Membrane Technology", *Water Technology*, 10(5):15-25, (August 1987).

24. Lykins, Jr., B. W. and Baier, J. H., "Point-of-Use/Point-of-Entry Systems for Removing Volatile Organic Compounds from Drinking Water", Significance and Treatment of Volatile Organic Compounds in Water Supplies, Lewis Publishers, (1990).

25. Eisenberg, T. N. and Middlebrooks, E. J., "Reverse Osmosis Treatment of Drinking Water", Stoneham, MA: Butterworth Publishers, (1986).

26. Baier, J. H., Lykins, Jr., B. W., Fronk, C. A., and Kramer, S. J., "Using Reverse Osmosis to Remove Agricultural Chemicals from Groundwater", *Journal American Water Works Association*, 79(8):55-60, (1987).

27. McClellan, S. A., "Membrane Types and Configurations", Proceedings of American Water Works Association Annual Conference, Membrane Preconference Seminar, Orlando, Florida, June 19, 1988.

28. Pohland, H. W., "Theory of Membrane Processes", Proceedings of American Water Works Association Annual Conference, Membrane Preconference Seminar, Orlando, Florida, June 19, 1988.

29. Reverse Osmosis for Home Treatment of Drinking Water, Michigan State University Cooperative Extension Service, Bulletin WQ24, January 1990.

30. Montemarano, J. and Slovak, R., "Factors That Affect RO Performance", *Water Technology*, 13(8):44-54, (August 1990).

31. Sorg, T. J. and Love, O. T., "Reverse Osmosis Treatment to Control Inorganic and Volatile Organic Contamination", U.S. Department of Commerce, PB84-223528, National Technical Information Service, Springfield, VA, (1984).

32. Cartwright, P. S., "Demystifying RO Water Treatment", *Water Technology*, 5(5):20-23, (August 1982).

33. Water Treatment Plant Design For The Practicing Engineer, edited by R. L. Sanks, Ann Arbor Science Publishers, Inc., Ann Arbor, Michigan, (1978).

34. Cartwright, P. S., "An Overview of Fine Filtration Technology", Pharmaceutical and Cosmetic Equipment, March 1985.

35. Shaparenko, G., "Home Water Distillers", *Water Conditioning and Purification*, 28(4), (May 1986).

36. Distillation For Home Water Treatment, Cooperative Extension Service, Michigan State University, East Lansing, Michigan, January 1990.

37. Survey of Test Protocols for Point-of-Use Water Purifiers, Department of National Health and Welfare, Ottawa, Canada, August 1977.

38. The Methane Gas Predicament, *Water Technology*, 11(8):42-46, October (1988).

39. Bellen, G., Anderson, M., and Gottler, R., "Defluoridation of Drinking Water in Small Communities", U.S. EPA Cooperative Agreement No. R809248010, Cincinnati, Ohio, 1985.

40. Laboratory Testing of Point-of-Use Ultraviolet Drinking Water Purifiers, Health and Welfare, Ottawa, Canada, April 1979.

41. Carrigan, P., "Water Disinfection Using Ultraviolet Technology", Proceedings of the Water Quality Association Annual Convention, San Antonio, Texas, March 14-18, 1990.

42. Wysocki, A. M., "Applications for Ultraviolet", *Water Conditioning and Purification*, 30(4), (May 1988).

43. Rice, R. G., "Ozone for Point-of-Entry/Point-of-Use Applications", Proceedings of the Water Quality Association Annual Convention, Atlanta, Georgia, March 4, 1989.

44. Wagenet, L. and Lemley, A., "Chlorination of Drinking Water", Cornell Cooperative Extension, New York State College of Human Ecology, Fact Sheet 5, September 1988.

45. Laboratory Testing and Evaluation of Iodine-Releasing Point-of-Use Water Treatment Devices, Department of National Health and Welfare, Ottawa, Canada, December 1979.

46. Montemarano, J., "Demand-Release Polyiodide Disinfectants", *Water Technology*, 13(8):65-67, (August 1990).

CHAPTER 3

APPLICATION OF POU/POE DEVICES

INTRODUCTION

The application of POU/POE can be as varied as there are models available. Historically, the POU/POE treatment industry is not new. POU/POE has been traditionally used for non-health related water impurities such as taste, odor, color, turbidity, iron, hardness, etc. in the home as well as for industrial and commercial applications. With the detection of toxins and carcinogens found in public supplies and individual wells, new technologies and applications have been developed. This chapter outlines the ways in which POU/POE devices are applied in the home for State and Federal hazardous waste actions, military applications, and commercial and industrial uses.

HOME USE

Two issues develop with homeowner use of POU/POE devices. The first deals with homeowners already on community or public water supply. There are only a very few public water supplies that do not meet the current safe drinking water standards and therefore most home applications are for individual sensitivities regarding the aesthetic aspects or to reduce real or perceived consumer risks from the water. The second issue deals with the discovery of harmful substances in drinking wells. Homeowner options tend to become a local issue with solutions often required by local officials (county health officers, town government officials, mayors or councilmen). The long-term institutional implications of these two situations are explored in detail in Chapter Seven. In either case, the homeowner has several options regarding the types of POU/POE devices that can be installed.

The individual technologies and their capacity for removing specific contaminants was discussed in Chapter Two. Those technologies can be applied in the home in different ways. The units can be distinguished by their use and location as shown below.

      (a) countertop
      (b) faucet-mounted
      (c) in-line
      (d) line bypass (third tap)
      (e) whole house

An illustration of the above (a through d) are shown in Figures 1 through 6. Whole house POE units can be installed in out-buildings, garages, or in basements. Closets have even been utilized. Figure 7 presents a generalized representation of a GAC POE unit installation.

With a countertop system, water can be diverted from the tap using an adapter or poured directly into the device from a separate container. Faucet-mounted units normally do not provide the contact time needed for organics removal and should only be used for taste and odor control (Figures 1 and 2). Figure 3 shows a portable countertop unit. The in-line unit (Figures 4 and 5) are normally connected to a single water faucet, usually under the kitchen sink, to treat all the cold water on that tap. The line bypass unit (Figure 6) provides an additional water faucet at the sink where the drinking water is obtained. By treating only the water used for drinking, the life of the POU device is extended. The POE treats all the water entering the house, is larger and more complex than POU devices, and has a higher initial cost.[1]

A database comprised of nearly 400 organizations was obtained by the USEPA and presented at a conference on POU devices. This data base provided the following snapshot of the industry.[1]

- 213 responses, 176 no responses
- 120 make or sell POU/POE devices
- 32 sell parts only (pumps, tanks, etc.)
- 13 state health boards
- 3 testing labs
- 23 returned letters (out-of-business)

Figures 8 and 9 show the number and type of devices available based on the survey. Table 1 lists the combinations of technologies offered to homeowners. The majority of the POU models in the database are in-line units (c). The RO units tend to utilize a line bypass (d) application more often than other technologies because of the amount of water wasted.

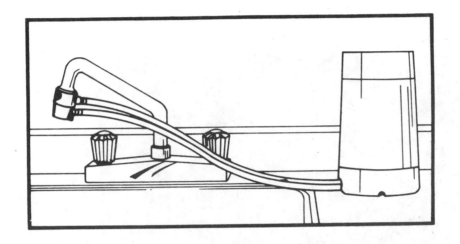

Figure 1.   Countertop Unit

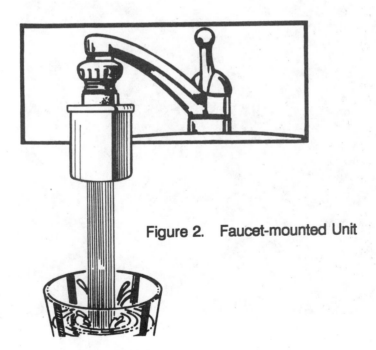

Figure 2.   Faucet-mounted Unit

Figure 3a.   Portable Unit

Figure 3b.   Interior of Portable Unit

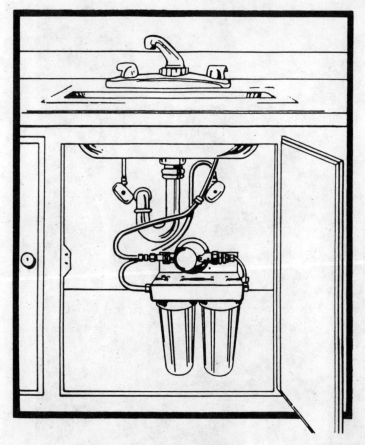

Figure 4.   In-line Unit

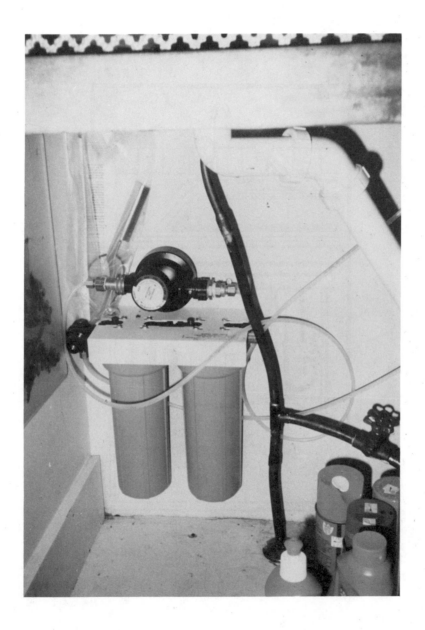

Figure 5.   In-line Unit with Automatic Shut-off

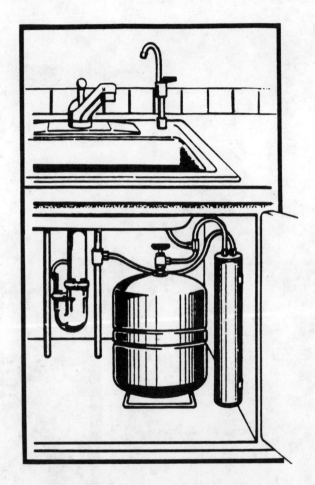

Figure 6.   Line By-pass Unit

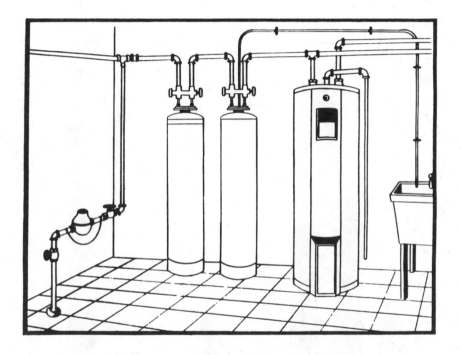

Figure 7.   GAC Whole-house Unit

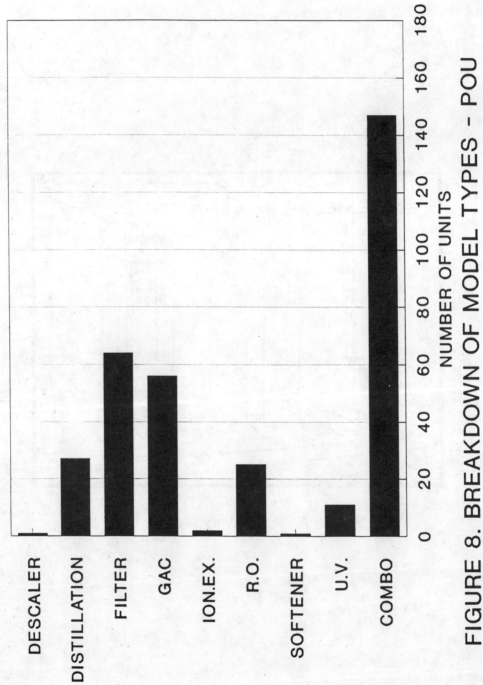

FIGURE 8. BREAKDOWN OF MODEL TYPES - POU

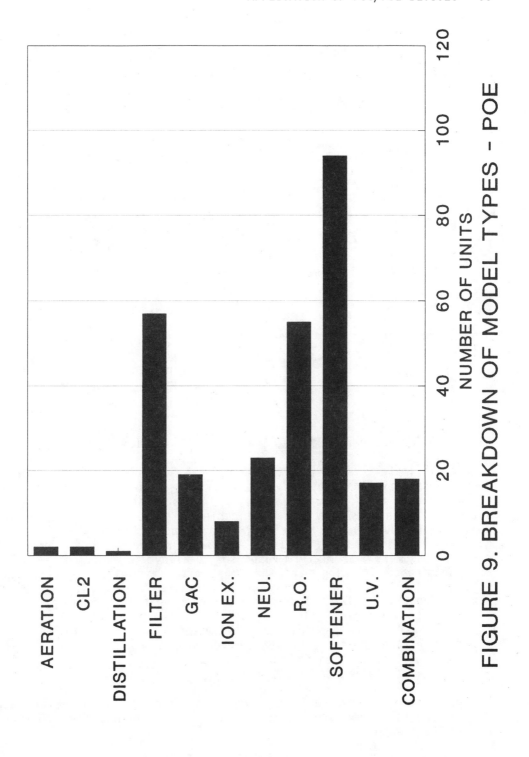

FIGURE 9. BREAKDOWN OF MODEL TYPES - POE

TABLE 1.   COMBINATION TECHNOLOGIES

TECHNOLOGIES	RECORD COUNT
DIS/FIL	9
DIS/GAC	1
DIS/GAC/RO	5
FIL/GAC	25
FIL/GAC/IEX/RO	1
FIL/GAC/RO	47
FIL/GAC/RO/UV	1
FIL/GAC/SOF	1
FIL/GAC/UV	1
FIL/RO	10
FIL/SOF	2
GAC/RO	6
GAC/UV	1
NEU/SOF	1

DIS - Distillation   FIL - Filtration
GAC - Granular Activated Carbon   RO - Reverse Osmosis
IEX - Ion Exchange   UV - Ultraviolet
SOF - Softening   NEU - Neutralization

STATE AND FEDERAL HAZARDOUS WASTE REMEDIATION APPLICATIONS

Approximately two-thirds of the Federal Comprehensive Environmental Resource Conservation Liability Act Superfund actions to date, deal with a contaminated drinking water supply. In most cases, the Safe Drinking Water Act standards are the applicable or relevant and appropriate requirements (ARARs) for Superfund remediations. States often want the Maximum Contaminant Level Goals to be used in order to more completely protect consumers. Short-term responses to alleviate an immediate danger of contaminated drinking water has most often been the use of bottled water. Alternative measures commonly considered during planned removal actions (non time-critical) or remedial investigations include:

- continue providing bottled water
- provide an alternative water supply
- installation of POU
- Installation of POE

Permanent remedies typically consider connection to a community water supply, provide a new water source, or install and maintain an individual POU or POE treatment unit.

Several Superfund sites have been employing both POU and POE devices for a few years. Various states have been utilizing POU/POE. In POU/POE applications, the predominant contaminants being removed are the chlorinated solvents including trichloroethylene, tetrachloroethylene, 1,1,1-trichloroethane, 1,2-dichloroethane, and trans-1,2-dichloroethylene.[2] Also being treated are waters contaminated by petroleum products, aldicarb, ethylene dibromide (EDB), or radon.[3] Table 2 summarizes the contaminants of concern and their influent levels. The removal efficiencies provided by the various systems range between 86 and 99+ percent. Table 3 presents an indication of the occurrences of contaminants at various concentrations for groundwater being treated at Superfund sites.

TABLE 2. SUMMARY OF EXISTING DATA
POE WATER TREATMENT STUDY[2]

SITE NAME & LOCATION	POE SYSTEM	CONTAMINANTS	MAX. INFLUENT	NO. POE SYSTEMS INSTALLED
State of Maine	Diffused air stripping	Gasoline and No. 2 Fuel Oil	240,000 μg/L	100
State of Maine	Diffused air stripping, or packed tower	Radon	400,000 pC/L	NA
Suffolk County, New York	Carbon cell	Aldicarb and daughter products	500 μg/L	4,500
Cattaraugus County, New York	2 Carbon Cells	TCE[a]	3,600 μg/L	37
Green County, New York	2 Carbon Cells	PCE[b]	79,500 μg/L	6
Onendaga County, New York	Packed Tower	TCE 1,2-DCE[c] 1,1-DCA[d]	690 μg/L	5 2 2
York County, Pennsylvania	Prefilter, Carbon cell, UV light	TCE PCE Carbon Tetrachloride 1,2-DCB[e]	4,600 μg/L	6

TABLE 2.   SUMMARY OF EXISTING DATA
POE WATER TREATMENT STUDY[2] (CONT.)

SITE NAME & LOCATION	POE SYSTEM	CONTAMINANTS	MAX. INFLUENT	NO. POE SYSTEMS INSTALLED
Berks County, Pennsylvania	Prefilter, Carbon cell, UV light	1,2-DCE TCE PCE DCA 1,1,1-TCA[f]	1,700 μg/L 23,000 μg/L 1,000 μg/L 50 μg/L 570 μg/L	28
Adamstown, Maryland	Prefilter, Carbon cell, UV light	TCE 1,1,1-TCA 1,1-DCA 1,2-DCE	520 μg/L 44,000 μg/L 210 μg/L 570 μg/L	18
Monroe County, Pennsylvania	Prefilter, Carbon cell, UV light	TCE 1,2-DCE PCE	7,000 μg/L 290 μg/L 30 μg/L	22
Florida	2 Carbon Cells	Naphthalene Total hydro-carbons Benzene Ethylbenzene 1,2-DCA[g] Toluene Xylene	12 μg/L 220 μg/L 210 μg/L 38 μg/L 89 μg/L 8 μg/L 63 μg/L	11
Polk and Jackson Counties, Florida	Prefilter, 2 Carbon Cells, UV light	Ethylene dibromide (EDB)	800 μg/L	850
Byron, Illinois	Prefilter, 2 Carbon Cells	TCE PCE	500 μg/L 130 μg/L	10

TABLE 2.   SUMMARY OF EXISTING DATA
POE WATER TREATMENT STUDY[2] (CONT.)

SITE NAME & LOCATION	POE SYSTEM	CONTAMINANTS	MAX. INFLUENT	NO. POE SYSTEMS INSTALLED
Elkhart, Indiana	Prefilter, 2 Carbon Cells, Packed Tower Aeration	TCE Carbon tetra-chloride	5,000 μg/L 7,500 μg/L	60 1
Uniontown, Ohio	Packed Tower Aeration	Vinyl Chloride Chloroethane	7 μg/L 2 μg/L	9

(a) - Trichloroethylene
(b) - Tetrachloroethylene
(c) - trans-1,2-dichloroethylene
(d) - 1,1-dichlorethane
(e) - 1,2-dichlorobenzene
(f) - 1,1,1-trichloroethane
(g) - 1,2-dichloroethane

TABLE 3.   INFLUENT CONTAMINANT OCCURRENCES AT
INDIVIDUAL HOUSEHOLDS INFLUENCED BY SUPERFUND SITES[2]

Contaminant	<100 μg/L	101-1000 μg/L	1001-5000 μg/L	>5000 μg/L
Trichloroethylene	85	60	17	3
Tetrachloroethylene	35	8	1	1
1,1-Dichloroethane	19	0	0	0
1,1,1-Trichloroethane	13	16	3	1
1,2-Dichloroethylene	8	7	1	0
1,1-Dichloroethylene	5	0	0	0
Vinyl chloride	9	0	0	0
Chloroethane	9	0	0	0
Carbon tetrachloride	8	10	0	0

*   The number of occurrences represents the highest concentration of a contaminant at an individual household.

Granular Activated Carbon Systems

The POE GAC filter systems currently being applied at Superfund sites usually consist of either a single carbon unit or two units connected in series. Some of the systems are also equipped with prefilters to remove particulates, water meters to measure flow, and ultraviolet (UV) light to disinfect the water after carbon treatment. The two-carbon-unit systems generally have sampling ports before the first carbon filter, between the two filters, and after the second filter.[2]

Most of the GAC treatment systems applied at Superfund sites are of the two-filter design. The second filter acts as a backup for the first; if the first filter should reach breakthrough, the second filter will provide continued treatment and protection. When filter replacement is necessary, the first filter is usually removed, the second filter is moved to the primary position in the series, and a new filter is placed in the secondary position. This procedure permits optimal use of the effective life of the activated carbon and provides satisfactory treatment of the contaminated water. (Disposal of saturated carbon is discussed in Chapter 4).

A single-filter system does not provide a backup filter; therefore the replacement scenario is different. In Suffolk County, New York (See Chapter 6), where single-carbon-filter systems were used, the systems are equipped with an automatic backwash function to prevent the filter from being clogged by particulate and/or bacterial buildup. Quarterly monitoring of the systems is required to detect breakthrough. In Florida, where several single-filter units are in use, the effective carbon life is estimated based on waterflow and expected contaminant loading, and the filters are replaced on a regular basis before breakthrough is expected to occur. Water monitoring is conducted on a quarterly basis to verify the security of the system.

At the Olean Superfund site in Olean, New York approximately 30 dual carbon filter systems have been installed to treat water contaminated by TCE at levels as high as 3,600 μg/L. The system typically consists of two GAC columns in series preceded by a particulate filter and water meter. Sampling ports are located before the first GAC filter, between the two filters, and after the second filter.

The POE GAC systems identified at Superfund sites in Region 3 consist of a prefilter, a flowmeter, two carbon cells in series, and a UV light. Each carbon cell contains 1.5 to 2.0 ft^3 of granular activated carbon. Where extremely high levels of VOCs exist (e.g., 44,000 μg/L TCE), several of the systems have a POE packed-tower air stripper connected to the beginning of the treatment train. The air stripper removes most of the VOCs from the influent water and thereby extends the effective life of the carbon filters.

The Region 3 communities currently using POE units are being administered by Federal personnel under emergency response legislation. However, this "emergency" has now been underway in some cases for more than three years. The communities did not rank high enough to be placed on the National Priority List (NPL) and thus become Superfund sites eligible for remedial action. This has created a situation where the federal personnel do not have a mechanism to turn over responsibility for the POE units. A variety of POE configurations exist as demonstrated by the photographs shown in Figures 10 through 13. It is not known who installed (Figure 13) the iodine disinfection unit or for how long it had been in service. Iodine units are not accepted for disinfection by the Pennsylvania Department of Health.

The residential filtration systems being used in Florida to remove EDB consist of two GAC units in series. These systems are installed between the existing pressure tank and the tap. A sediment filter and a water meter are installed in front of the carbon units. The sediment filter, a disposable cartridge, removes any particles larger than 5 micrometers to prevent the carbon units from clogging. The water meter permits accurate measurement of the water passing through the units. A UV light is installed after the second filter to disinfect the treated water. Sampling ports are placed before the sediment filter and after each adsorption unit.

Experience at the Elkhart, Indiana Superfund site has demonstrated the risk of extrapolating GAC isotherm tests to predict breakthrough. In two years, only 1 of the 54 units has experienced breakthrough although the isotherm tests predicted much earlier breakthrough.

Each adsorption tank at Elkhart is 10 inches in diameter and 54 inches high and is made of polyester fiberglass. Each tank contains 1.5 ft^3 of granular activated carbon. The remaining volume is occupied by a PVC distributor and gravel support media. The tanks are connected by PVC piping, and all valves are either brass or hard copper.

The adsorption system is designed for a maximum flow rate of 5 gallons per minute (gpm), which provides a real contact time of 2.2 minutes. A fully opened faucet allows a flow rate of approximately 3 gpm, which provides an effective contact time of 3.7 minutes.[3]

Air Stripper Systems

Two aeration methods typically applied in POE treatment are packed-tower aeration and diffused bubble aeration.

Figure 10.   Dual GAC POE Unit Installed for a Mobile Home

Figure 11.   Dual GAC Unit with Steel Tanks

Figure 12.   Single GAC Unit with Softener

Figure 13.   Pre-disinfection with Iodine on Same Unit in Figure 12

POE Packed Tower Aeration

POE Packed Tower Aeration (PTA) units in EPA Region 5 are generally designed for 90 percent removal of TCE at a 5 gpm flow rate. This design appears to effectively remove other volatile organics of concern.   POE PTA units have been installed at Superfund sites in Pennsylvania, Ohio, Indiana, and Maryland. Preliminary results indicate that 93 percent of TCE at 4,000 μg/L, 89 percent of TCA at 800 μg/L, and 99.9 percent of Vinyl Chloride at 5 μg/L can be removed. Overall efficiency at high VOC levels (> 1000 μg/L) are in the 80 percent range because the air becomes saturated with VOCs thus limiting transfer of the VOCs from the water to the air. Figures 14 and 15 describe a typical POE PTA unit. Figures 16 and 17 show a PTA unit in series with a softener at an Akron, Ohio Superfund site. POE PTA is currently installed in-series with GAC units under basement stairways in two homes in Elkhart, Indiana. Figure 18 displays this arrangement. Four feet of packing material removes approximately 80 percent of carbon tetrachloride and TCE at 7,500 μg/L and 5,000 μg/L respectively, before final GAC adsorption.

Diffused Bubble Aeration

There were two different types of diffused bubble aeration systems that were used originally for VOC removal but they are now being used for the removal of Radon (Rn). One system uses a multi-staged vessel to approach plug flow efficiency.[4]   Figure 19 is a schematic similar to a system located in Derry, New Hampshire being used for Rn removal. The second system uses a shallow tray single stage approach with a very high air to water ratio. Both systems are capable of Rn removal of greater than 99 percent. They also avoid the build-up of Gamma activity associated with GAC and the question of disposal problems. Figures 20 and 21 describe the shallow tray aerator unit.

Other Federal Applications

The National Park Service[5], Department of Defense,[6,7,8,9] and State Department are responsible for providing potable drinking water under various conditions. These include visitor centers, campgrounds, and restaurants to personal treatment devices for diplomats traveling and residing overseas and for military bases and operations in the field.

Treatment units evaluated under these conditions have included (1) cartridge filters for turbidity and *Giardia lamblia* cyst removal at Mount Rainier National Park, (2) reverse osmosis and ultrafiltration membranes, and conventional filtration for virus and possible biological agents used in warfare, and (3) GAC and air stripping for solvent and fuel contaminated aquifers at military installations.

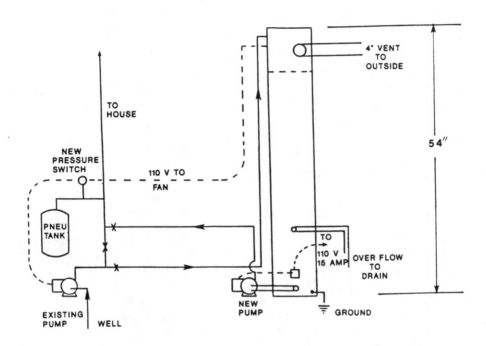

Figure 14. Packed Tower Aeration Unit Schematic

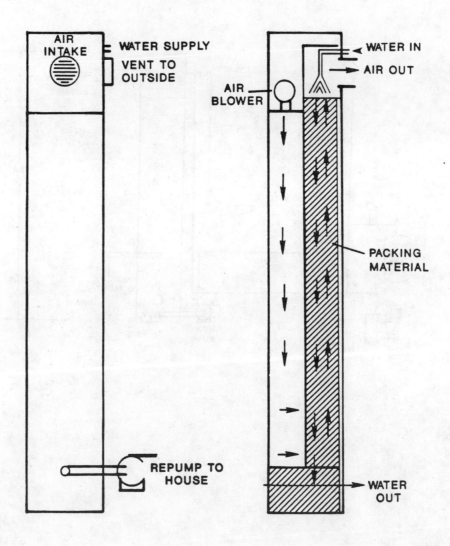

Figure 15.   Detail of POE Packed Tower Air Stripper

Figure 16. Softner Installed In Closet

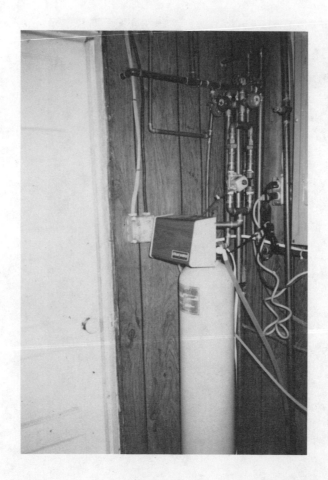

Figure 17.   Prefilter and Packed Tower Air Stripper

Figure 18.   Packed Tower Air Stripper In-series with Dual GAC Tanks

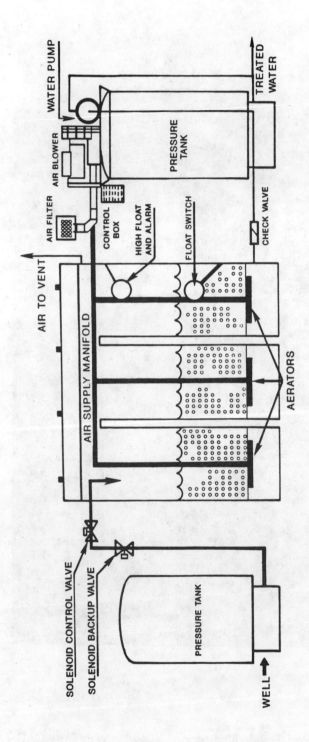

Figure 19.   POE Diffused Bubble Aeration System Schematic

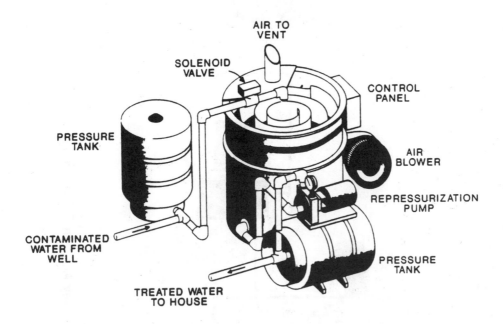

Figure 20.   Shallow Tray Aeration Unit

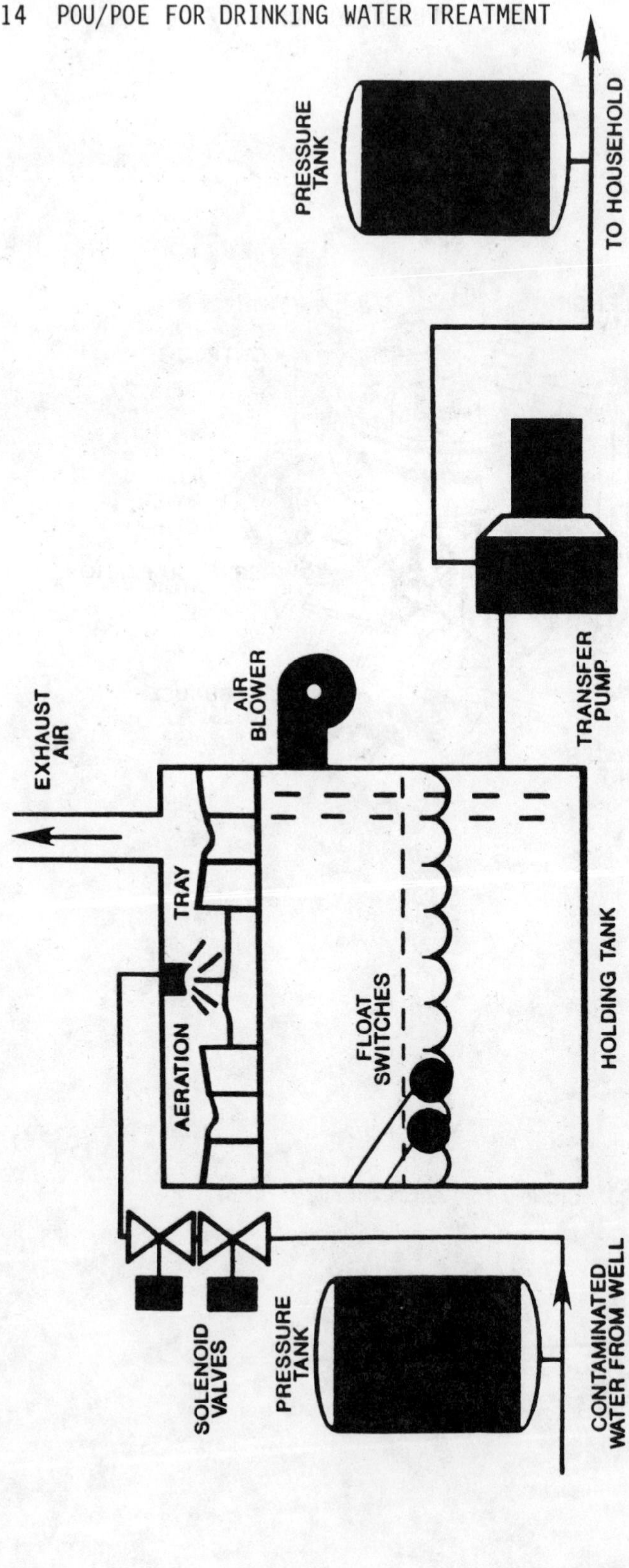

Figure 21. Shallow Tray Aeration Schematic

National Park Service

The National Park Service (NPS) operated 23 surface water supplies in the State of Washington in the mid 1980's. Only one had more treatment than disinfection. The other 22 supplies varied in operation from seasonally serving eight faucets in a picnic areas, to Longmire, serving an administrative area with about 60 permanent year-round residents, a visitor center, hotel, restaurant, and overflow campground.

The watersheds of these 22 supplies are owned by the NPS except for one which is owned by the U.S. Forest Service and one which has a small portion on Forest Service lands. Consequently, there is no agriculture or logging, no septic tank systems, no treated sewage discharges, and no pesticide use on the watersheds. None of the supplies are obtained from glacial streams.

Raw water turbidities are less than 1 NTU except for periods of heavy rainfall or snowmelt. These events usually occur during the early winter and in the spring when visitor use and water demand is low and most of the seasonal supplies are out of service. Turbidities of 5 NTU are seldom reported, although turbidities of 45 NTU were reported several years ago from a since-abandoned supply and 30 NTU was once reported from an active supply. Except for coliform organisms, raw water meets the National Interim Primary Drinking Water Regulations.

Because the Park Service has a policy to eliminate surface supplies that are only disinfected wherever possible, the Pacific Northwest Region has abandoned several surface and groundwater sources. Cartridge filtration was investigated as an alternative to conventional treatment plants for those supplies where groundwater is not available and conventional treatment is not feasible because of size and/or length of season.

Cartridge filtration, as a supplement to disinfection, began on the Longmire supply (Mount Rainier National Park). This was a field test of the feasibility of operating the equipment. Since then cartridge filtration has also been installed on five other NPS supplies.

Longmire. The Longmire intake is an eight feet high, concrete, overflow dam on an unnamed creek. The stream slope is quite steep and the irregular impounded pool is approximately 15 feet long and 10 feet wide. The intake is a 4-inch Teed line with two screened pipes suspended at about mid-depth of the pool about three feet from the right bank. The intake is about 250 feet in elevation above the filter installation.

The cartridge filters used at Longmire performed satisfactorily. The average volume filtered per cartridge used was 34,000 gallons for 50-micron cartridges, 112,600 gallons for 25-micron cartridges, and 181,140 gallons for 5-micron cartridges. Turbidity reductions were about 25% with higher percent removals at higher turbidities. For example, typical removals were 10% at raw water turbidities of 0.2 NTU and 50% at raw water turbidities of 1.8 NTU. The filters did not, however, consistently produce water with 0.2 NTU or less turbidity except when raw water turbidities were only slightly above that value.

The Longmire data tend to show that the 50-micrometer filter provided most of the turbidity removal. The 25-micrometer filter provided backup protection for the 5-micrometer filter, whose role was Giardia removal. There was some indication that some of the dirt caught on the 50-micrometer filter came through and was removed on the 5-micrometer filter.

For the 50-micrometer cartridges in FY 1983, the minimum volume filtered per cartridge was 4,500 gallons, the average was 34,000 gallons, and the maximum was 167,500 gallons. The daily log sheets show that the turbidity was not significantly high. It ranged from 0.41 to 0.6 NTU. Data indicates that consistent, moderately high turbidity (that nevertheless meets the turbidity MCL) will shorten the filter runs more than an occasional spike up to 1.2 to 2 NTU from base turbidities of 0.2 to 0.3 NTU.

The field study resulted in the following conclusions:

1.   Cartridge filtration should be used with caution to reduce turbidity to meet the maximum contaminant level of 1 NTU. The data from this field test suggest they might be effective, but that manpower and cost of supplies would make them expensive to operate. When turbidities are consistently greater than 1.5 NTU, cartridge filtration is most likely to be economically feasible only for small (less than 1,000 gpd) supplies where dollars per day is more important than dollars per one thousand gallons.

2.   The cartridges used in this field study did not rupture and release a slug of dirt into the water, when stressed by high turbidity. On one occasion they clogged completely but did not fail mechanically. They did produce water with more than 1 NTU when raw water turbidity was 3 NTU.

3.   Experience demonstrated the adverse effects of high (4 NTU and above) raw water turbidity on filtered water turbidity and cartridge consumption. When these events are infrequent and of short duration, they may be tolerable, particularly if there is sufficient storage to eliminate the need to treat water during the adverse water condition.

4.   The use of disinfection and three sizes of filters in series,
     e.g. 50-micrometer, 25-micrometer, and 5-micrometer (of
     demonstrated ability to remove cyst-sized particles) is a
     satisfactory substitute for chemical coagulation-filtration to
     (a) produce water meeting physical and microbiological maximum
     contaminant levels and (b) remove _Giardia_ cysts when raw water
     turbidities are typically 1 NTU or less.

U.S. Army.   Valcik et al., surveyed 36 Army installations across the
country with a total population of 575,000 persons.[7]   Military drinking
water quality problems have mirrored the temporary and long-term health
and aesthetic problems of civilian supplies.

     Alternatives to providing a more acceptable quality of drinking water
have included central-system improvement (e.g., modifications in the water
treatment process and flushing of the distribution system), use of bottled
water, boiling, hauling water, quick connection to nearby community
sources, and POU/POE treatment technology.   In some cases POU/POE
treatment technology has been judged to be the most viable available
option. Several illustrative cases of POU/POE technology application will
be presented.

     The size of the installations surveyed ranged from 500 to 48,200
persons with a median population of 18,300 persons on an effective
population basis. Resident populations were as small as 19 persons and
total populations (residents plus non-residents) were as high as 60,000
people.   Water supply was obtained from both surface and groundwater
sources, purchased from a supplier, or treated by the installation before
distribution via an Army-owned central system.

     Table 4 provides a summary of the survey regarding the use of POU/POE
devices. Of the 36 installations surveyed, 24 installations (2/3) were
reported to use POU/POE technology. Essentially all devices are employed
for treating water from a distribution system supplied by a central water
treatment facility.

     The survey showed that the total number of devices reported in use
was 768 of which 85 percent were of the POU variety. The largest number
of devices used at any one installation (300) were also of the POU variety
to remove turbidity for drinking purposes (Installation 9, Table 5). A
wide variety of manufacturers' devices and models were found to be in use,
and only 184 (24 percent) of the devices were reported to be approved or
certified by a recognized entity such as the National Sanitation
Foundation (Table 5). Such devices are the result of user choice and
action. Unit costs, where reported, ranged from approximately $40 for a
POU combination particulate/adsorption filter to $20,000 for a large POE,
ion exchange unit installation.

TABLE 4.   SUMMARY OF POU/POE DEVICES AT ARMY INSTALLATIONS[*]

	Installations	No. of Devices			Treatment Technology	POU/POE Treated Water Uses
		Total	POE	POU		
Health-related uses	3 (8%)	28 (4%)	25	3		
- Bacteriological safety- at various non-CS wellwater sites	3	25	25	0	Disinfection (chlorination)	Drinking and other household-type uses
- Hemodialysis - CS situations	2	3	0	3	RO, IEX, AF	Hemodialysis (hospitals)
Improved aesthetic quality	11 (31%)	587 (77%)	33	554		General household uses or specific uses - i.e. drinking, beverages, icemaking, showering
- Better physical character- istics (color, turbidity, T/O) in CS situations	8	563	10	553	PF, AF	
- Reduced hardness/TDS - in CS situations except one individual wellwater case	4	24	23	1	IEX	
Enhanced medical/industrial water quality - in CS situations	13 (36%)	153 (21%)	62	91		
- Removal of discoloration/ turbidity	3	29	5	24	PF	Film processing, metal finishing
- Reduction of hardness/TDS	9	120	57	63	IEX	Medical/lab uses, boilers, laundry
- Removal of most contaminants	6	8	0	8	Distillation RO	Medical/lab uses
TOTAL		768	120	648		

CS - central system
RO - reverse osmosis
IEX - ion exchange
T/O - taste/odor
TDS - total dissolved solids
PF - particle filtration
AF - air filtration

* Of 37 installations surveyed, 24(67%) use some form of POU/POE device.

TABLE 5. SPECIFIC APPLICATIONS OF POU/POE DEVICES AT ARMY INSTALLATIONS

Installation	Location/ Activity	Type of Service	Treatment Technology	Reason for choice - to reduce/eliminate	Product water use	Number of devices	Unit cost, $
1	Hospital - labs	POU	IEX	TDS	I-lab	44	
		POU	IEX, PF	TDS, turbidity	I-photo	2	
		POU	PF	Turbidity, discoloration	I-photo, laser cooling	15	$11,000
	Industrial building	POU	RO w/IEX, AF	Microbes, inorganics	H-dialysis	2	
		POU	PF	Discoloration	A-drinking	1	
		POU	IEX		I-steamer	1	
		POU	IEX	TDS	I-boilers	2	
2	Power plant Hospital - dialysis	POU	RO	Physical, chemical & microbiological	H-dialysis	1	3,000
3,4,5	Hospital - dining area	POU	IEX	TDS	A-culinary, drinking	1	4,400
	- eating areas	POU	PF, AF	Turbidity, T/O	A-culinary, ice, beverages	7*	40
	Administrative areas	POU - F, USLB	AF	Discoloration, T/O	A-drinking	5	150
6	Housing	POU-USLB	AF	T/O (chlorinous)	A-culinary, drinking	----	
	Administrative areas	POU	AF	"Personal choice"	A-drinking	----	
7	Remote locations - petrol transfer site, golf clubhouse	POE	C	Batteries	H-drinking	4	374
8	ski lodge, salvage yards	POE	C, IEX	Bacteria, TDS	A, H-household	3	
	Research & development	POE	PF	Sediment, discoloration	I	5*	
9	Administrative, housing school, recreational areas	POU-USCT	PR	Turbidity	A-drinking, culinary	Estimate 300	

TABLE 5.   SPECIFIC APPLICATIONS OF POU/POE DEVICES AT ARMY INSTALLATIONS (CONT.)

Installation	Location/Activity	Type of Service	Treatment Technology	Reason for choice - to reduce/eliminate	Product water use	Number of devices	Unit cost, $
10	Power plants, (4), dental clinic	POU	IEX	Hardness	I-boilers, dental	9	$ 1,500 (dental)
	Shower facilities	POE	IEX	Hardness	A-showering	2	20,000
	Remote sites (17), field training and support	POE	C	Bacteria	H-drinking	17	1,000
	WTP-lab	POU	D	Hardness, metals	I-lab	1	6,950
	Dining, work, visitors quarters areas	POU	PF*	Sediment	A-ice	140	50
11	Hospital - labs	POU	IEX, D, RO, PF	Impurities	I-lab	1	
12	Power plants (30), laundry	POE	IEX	Hardness	I-boilers, laundry	31	
13	Quarters	POU	PF	Turbidity	A-household	1	
14	Power plants (13), laundry	POE	IEX	Hardness	I-boilers, laundry	14	
	Firing range	POE	C	Bacteria	H-drinking	1	
15	Hospital - lab	POU	D	Impurities	I-lab	1	
16,17,18	Hospital	POU	RO	Impurities	I-lab, pharmacy	1	
19	Hospital - surgery	POU	IEX	Hardness	I-medical	1	
	- power plant	POE	IEX	Hardness	I-boilers	2	3,700
20 21	Dining facilities (4) clothing sales, gym	POE	IEX	Hardness	A-drinking, culinary, showering	6	
22	Housing	POE	AF	Color	A-household	5	
23	Laundry	POE	IEX	Hardness	I-laundry	2	
24	Hospital	POU	IEX	Hardness	I-lab	2	
25,26	Housing	None	IEX	Hardness	A-household	2	

TABLE 5.   SPECIFIC APPLICATIONS OF POU/POE DEVICES AT ARMY INSTALLATIONS (CONT.)

Installation	Location/Activity	Type of Service	Treatment Technology	Reason for choice - to reduce/eliminate	Product water use	Number of devices	Unit cost, $
27	Dental clinic	POU	IEX, RO	Impurities	I-dental	2*	
28	Power plants (5)	POE	IEX	Hardness	I-boilers	5	$ 1,500 to 10,000
29,30	Medical facilities	POU-USLB	D	TDS	I-lab	2	
31		POE	IEX	Hardness	I-lab	1	
		None					
32,33	Dental clinic	POE	IEX	Hardness	I-lab	2	2,995
		None					
34	Photo lab	POU	PF	Turbidity, Discoloration	I-lab	1	
35	Housing	POU	PF	Turbidity, Discoloration	A-household	35*	
		POE	PF	Turbidity, Discoloration	A-household	5	
	Industrial	POU	PF	Turbidity, Discoloration	I-plating, photo processing	6	
	Workplaces - water coolers	POU	PF/AF	Turbidity, Discoloration	A-drinking	60	
	Workplaces - tap	POU	PF	Turbidity, Discoloration	A-consumption	4	
36	Housing	POE	IEX	Sulfide, iron	A-household	10	1,160
	Power plants	POE	IEX	TDS	I-boilers	----	

ABBREVIATIONS: POE = point-of-entry, POU = point-of-use, F = faucet mounted, USLB = under-sink-line-bypass, USCT = under-sink-coldwater-tap,
IEX = ion exchange, PF = particulate filtration, RO = reverse osmosis, GAC = adsorption filtration, C = chlorination,
D = distillation, TDS = total dissolved solids, T/O = taste/odor, A = improved aesthetic value, H = health-related, I = industrial.

* Certified by National Sanitation Foundation (NSF)

Most of the POU/POE devices were used for improvement in the aesthetic quality of centrally supplied water. Over 3/4 of the total, or 587 were in this category; 94 percent of these (554) were of the POU type. These devices were utilized to improve physical water quality characteristics (i.e., eliminate color, turbidity and taste/odor) via particulate and/or adsorption filters, and to reduce hardness/total dissolved solids (TDS) via ion exchange. These devices were located at various housing, administrative, workplace, and dining/mess hall areas at almost 1/3 (11) of the installations surveyed.

At just over 1/3 (13) of the installations surveyed, POU/POE treatment devices (about 1/5 or 153) were used for enhancing central-system water quality for a variety of medical and industrial purposes. About 40 percent (62) of these devices were POE, while about 60 percent (91) were POU. These devices were used to: remove color and turbidity via particulate filtration, reduce hardness/total dissolved solids via ion exchange in the majority of instances, and remove most contaminants via distillation and/or RO. Enhanced water quality was used for laboratory and dental purposes, in hospitals, film processing, laundry washing, and boiler applications at medical and laboratory facilities, laundries, photo labs, and heating plants.

The following remarks on the use of POU/POE technology were made by health officers and water supply personnel as part of the survey.

- Such devices are not considered necessary because the army is currently meeting the water quality standards under current drinking water regulations.

- Some devices are considered to be potentially dangerous (which is an allusion to buildup of high and potentially harmful levels of bacteria in devices containing granular activated carbon or breakthrough of contaminants).

- One location's proposed regulation prohibited the use of these devices.

- Sales pitches by solicitors may convince consumers that local-treated water is not good thereby urging consumers to purchase unnecessary devices.

- Installation engineer authorities are opposed to consumers tampering with residential plumbing systems.

These remarks are not unexpected and echo the home residential applications situation.

Survey personnel reported an inability to determine the full extent to which devices are used in housing/quarters areas. The following three case studies illustrate the use of POU/POE at Army installations surveyed.

Case 1 - Inadequate Distribution System O/M. The installation is located in one of the Middle Atlantic states and typifies older moderately-sized community water systems. A pre-World War II conventional rapid sand fil-tration water treatment facility processes approximately 2 MGD of water from a stream. Lime, with token amounts of sodium silicate, renders the finished water generally non-corrosive (based on the Langelier Saturation Index). The finished water, in passing through the nearly 5-decade old, predominantly cast iron distribution system to the various points of demand, tends to become somewhat corrosive, especially in the lower-demand areas, and certainly in the dead-end mains. Consumer complaints of "rusty" or discolored water, especially after a long weekend (low-flow), reached a level that technical assistance was sought by the installation. Investigation revealed that POU/POE particulate/adsorption filters had been installed in numerous situations - virtually in all the individual residences, and on numerous water cooler water supply lines, beginning about four years prior to the date of the investigation. Servicing involved replacing filter cartridges in response to consumer calls of reduced output. These calls increased in intensity when the fire department tested fire hydrants causing the iron-containing sediments to be stirred up, and then turned off the hydrants before letting the water clear. Further investigation revealed a prime reason for the increasing prevalence of discolored water - the distribution system had not been effectively flushed for four years due to budgetary constraints. Thorough flushing now occurs at least annually as required by Army regulations. All the devices are still in place. As demonstrated by this case, an effective water distribution system flushing program can have a major role in minimizing undesirable aesthetic water quality. Therefore, one should not totally rely on POU/POE devices to solve aesthetic problems in commu-nity water systems.

Case 2 - Capitalizing on the Judicious Use of POE Devices. This case concerns a partially-active ammunition plant in the South with only 16 residents and about 1,700 employees. Due to its reduced level of produc-tion, only 0.5 MGD or about 20 percent of the available capacity supplied from several deep wells was used. Primary uses were industrial and fire protection; only about 0.5 percent of the capacity or 2,400 gpd (16 X 150 gpcd) was required by the residents for domestic purposes, and another 16 percent or 80,000 gpd (1,700 x 50 gpcd) was required by the employees for consumption and showering. The actual amount of water for drinking by employees was only about 0.2 percent or 850 gpd (1,700 x 0.5 gpcd). Unusual aspects of the water supply which was only chlorinated, were the undesirable aesthetic characteristics - faint hydrogen sulfide ("rotten egg" odor), substantial iron concentrations, and moderately corrosive water. By the time the water traveled through the primarily cast iron World War II vintage sprawling distribution system, which was underuti-lized from normal demand but at times reached points of demand, further deterioration of the water quality occurred. Iron concentrations in-creased due to the interaction of the corrosive water with the cast iron mains particularly in lower demand use areas resulting in discolored water. An associated difficulty in maintaining chlorine residuals throughout the distribution system gave rise to many coliform problems. For the residents, POE devices using a zeolite ion exchange process specific for hydrogen sulfide and iron removal were in place at each of the 16 residences. Contractor servicing/maintenance occurred about every 10 weeks. The most appropriate devices to be placed in the production areas to improve drinking water quality were determined. This case points up the advantage of POU/POE technology in situations where only a small fraction of water is needed for household use. The expense of treating all supplied water to an acceptable quality should be evaluated along with central-system actions like flushing a very extensive system on a frequent basis.

Case 3 - Use of POU/POE Devices for Fluoride MCL Compliance. At this installation in the Southwest, the community water system in the main post area with a total population of about 1,100 is served by chlorinated ground-water supplies. These supplies are characterized by a high, natural fluoride content of 1.7 to 3.2 mg/L and high mineral content of 1,100 mg/L total dissolved solids (TDS). Under the impetus of the National Interim Primary Drinking Water Regulations, an exemption was granted from the health-related fluoride MCL of 2.4 mg/L in 1979 to allow time to plan and design a central system to reduce fluoride in the drinking water to below the existing MCL. These efforts culminated in the installation of an electrodialysis reversal (EDR) water treatment facility and a separate potable water distribution system which became operational in 1987. At the beginning of the interim period, reverse osmosis POU technology was employed as the drinking water alternative of choice. About 50 reverse osmosis units were installed in the workplace to serve a workforce of 300. Rapid plugging of the membranes by the TDS required frequent membrane replacement which became an unmanageable servicing problem for the installation maintenance personnel. These units were removed from service by the end of 1982, and replaced by bottled water. Bottled water continued to be used in the housing areas with some 800 residents, when it was begun in 1979, until the EDR plant became operational in 1987. The cost for bottled water at about $125,000 per year was considered very high in relation to the cost for the central system which included $320,000 capital cost for the EDR plant. In the industrial areas which also have ground-water sources with high fluoride contents, a study has been initiated to explore alternatives to reduce fluoride below the newly mandated 4 mg/L MCL. A key element of this evaluation will include POU/POE treatment as a potable water supply alternative. This case highlights the necessity for careful evaluation of required maintenance for selected devices and insuring the availability of adequate manpower resources.

State Led Applications

Figure 22 shows an advanced oxidation POE unit used by the State of Connecticut to treat a well contaminated with PCE. The well serves four homes. The unit utilizes GAC following contact with ozone/UV in combination to ensure complete removal of the PCE. The ozone/UV process removed

Figure 22.   O$_3$/UV/GAC State Lead Site Remediation

an average of 89 percent of the raw water PCE levels originally found in the range of 300-600 $\mu g/L$. The GAC unit removed the remaining PCE. No ozonation by-products were detected.

COMMERCIAL, MEDICAL, AND INDUSTRIAL APPLICATIONS

Water purification applications in commercial, medical, and industrial sectors probably represent the largest potential for sales of POU/POE and package plant units. Yet, this remains the least penetrated market.[10] Reverse osmosis systems tend to compete with distillation, softening, and deionizing units in medical and industrial applications. RO can and is used in conjunction with GAC filtration in office, commercial, and light industrial activities in direct competition to bottled water.

Office

Applications within this category include employee drinking water needs in:

- offices
- businesses
- factories
- hospitals

Office employees oftentimes want an alternative to tap water--usually for one or more of the following reasons:

1) As a more palatable, enjoyable replacement for tap water.
2) As a more conveniently located source of water in the facility.
3) As a source of chilled and boiling hot water--particularly when no kitchen facilities exist.
4) As a legal requirement to furnish employees with a supply of drinking water.

Traditionally, the bottled water dispenser or cooler, supplying room temperature or hot/chilled water, can be found in almost every type of business or large commercial facility. It should also be noted that many offices simply have bubbler type drinking fountains for their employees. These are usually planned in the original construction of the building and are the least expensive means of providing drinking water. These can rarely be converted over to an alternate system incorporating water purification.

In areas of low total dissolved solids (<300 ppm) GAC filtration systems can be applied. When using RO one needs to consider the following in office/commercial applications.

1) There are certain types of cooler dispensers that are not suitable for dispensing reverse osmosis purified water because of their materials of construction. RO purified water is often too aggressive and can, over a period of time, deteriorate certain expensive cooler components. Historically, this has been one of the most common causes of customer dissatisfaction in the office market as well as being an expensive pitfall to dealers.

2) Avoid using coolers that store purified water at room temperature in an atmospheric container.

3) Avoid using copper lines for distributing purified water.

4) It is frequently easier, safer and cheaper to install an RO system close to the influent water and drain, and run one low pressure purified water line to the cooler dispenser(s).

5) If more than one cooler is to be installed at a given facility, the remote cooler approach is preferred. The RO system should be sized to meet the total needs of the customer. Each cooler dispenser will then receive as much or as little water as is needed at its location.

6) If activated carbon filtration systems are used in conjunction with cooler dispensers, they should be capable of 1 micron absolute removal to ensure the microbiological quality of the cooler reservoirs.

Commercial and Light Industrial Applications

Commercial and light industrial activities have used and are anticipated to increase their use of POU/POE devices to produce water for a variety of different applications such as:

- restaurants
- ice manufacturing
- beverage production
- water vending machines
- bottled water production
- car wash systems
- horticultural nurseries
- medical applications
- industrial applications
- farm applications

Restaurants

Traditionally, restaurants have not felt pressed to invest in a sophisticated purification system. However, as public awareness or perception of deteriorating water supplies increases, the restaurant industry will respond.

Many natural food restaurants, finer traditional food establishments and restaurants located in very bad water areas are beginning to portray a health conscious menu. They can provide an alternate source of water for their customers. Applying carbon filtration and/or RO to ice machines, beverage systems and steamers can be justified by a restaurant for several reasons:

1)   It provides a conveniently dispensed source of water for drinking and for other beverages such as coffee and juices;
2)   it is affordable enough to be used in the preparation of foods and soups;
3)   it is a practical means of reducing service on restaurant equipment such as ice machines, coffee makers, soda and juice dispensers and espresso machines.

One such example of this is the Alpha restaurant in Cincinnati, Ohio. The original owner was a former traveling salesman with definite ideas on how restaurants should treat their customers. As part of his overall plan to develop a restaurant that catered to its customers and provide a healthy ambiance, he decided to install GAC filters six years ago to treat the water used for making ice, cooking, beverages, and food preparation. This was mostly because of some mistrust of the chemical quality of the local drinking water that came from the somewhat industrialized Ohio River basin. Local water is used for dish washing. Customers who routinely use bottled water or have their own POU/POE units notice the water's lack of chlorine taste or different coffee taste. Whereas customers accustomed to the local water do not normally realize the difference. The owner uses the water treatment as part of his marketing strategy in that a certificate describing the POE system is prominently displayed next to the host/hostess stand along with favorable restaurant reviews from local newspapers. Photographs of the restaurant, and the certificate follow in Figures 23-25.

The Alpha uses four treatment processes in-series to improve the aesthetic quality of the local tap water which is already in compliance with the Safe Drinking Water Act MCLs. A five micron sediment pre-filter is followed by a five micron diatomaceous earth ceramic filter capable of removing *Giardia*, although none has been identified in the tap water. This is advertised as helping the residual chlorine remove bacteria. A wood carbon downflow filter follows the ceramic filter to remove residual chlorine. Last in the series of treatment is a coal based solid GAC block. Once a year the diatomaceous earth ceramic filter is scraped and cleaned while the other filters are removed and replaced for a $45 fee (1991). Figures 26 and 27 show a restaurant's system in Pennsylvania to remove TCE.

Figure 23.   POE Unit in Restaurant

Figure 24.   Water Treatment Certification Letter in Restaurant

Figure 25.   Restaurant Promoting Its Water Quality

Figure 26.   Portion of Restaurant's Elaborate Water Treatment Unit

Figure 27. Portion of Restaurant's Elaborate Water Treatment Unit

Ice Machines

Commercial ice machines are generally classified as either "flakers" or "cubers". Flakers work essentially like common ice machines. A preset volume of water fills a mold and is then frozen and dispensed to a holding bin. Cubers, on the other hand, cascade water over a mold surface for contact freezing, layer-by-layer. Although a more complicated piece of equipment, this process actually lowers the TDS of the water, producing clearer ice cubes in high TDS water supplies. Because of the method of ice production, cubers use about twice as much water to make ice.

Ice machines can be treated with carbon/mechanical filtration -- the largest application -- and reverse osmosis. Chlorinated water and common rust and sediment can spoil any beverage in a food service establishment. Most food service carbon system manufacturers use 0.65 to 5 micron mechanical filtration, a bed of granular activated carbon and a scale inhibitor to control carbonate scaling. Systems are prepackaged and include pressure loss gauges to signal cartridge changes.

Reverse osmosis is usually applied only to flakers in high TDS areas (>500 mg/L) or where the TDS is doing damage to the equipment. It is important to carefully size the RO production rate and storage/make-up capacity so the ice machine does not run out of water. Carbon post-treatment is necessary on the RO to control taste/odor problems.

Beverage Systems

The most common beverage systems that benefit from treatment are post mix machines and coffee makers. Post-mix machines are beverage dispensers that combine syrup and carbonated water to produce the desired product. The ratio of carbonated water to syrup is critical to both the taste of the product and the profitability of the post-mix machine. In water supplies with high levels of chlorine, chloramines, chlorophenols and other taste/odor causing compounds, the typical remedy is to increase the syrup concentration to add sweetness to the beverage taste.

A carbon system on the post-mix machine will produce many more glasses of soda from the same syrup container. Since the bulk of the cost is the syrup, the carbon system pays for itself and the product tastes better. Mechanical filtration ranges from 0.65 to 5 microns, and pressure loss gauges indicate the need for cartridge changes due to particulates. The gallonage is approximate and fluctuates according to local conditions. Totalizers are not usually used. When the carbon is exhausted, there is a noticeable change in the quality of the product.

Automatic feed coffee makers are another excellent possibility for water quality improvement. Coffee is 98 percent water, the poor quality of which commonly ruins the brew. Scale-causing minerals and high TDS water can also damage equipment. Although water treatment somewhat increases costs, improved coffee taste is the basis for the sale.

The choice of carbon or RO depends on the TDS and any scale problems experienced. Generally, if the TDS is less than 500 mg/L, a food service carbon system with scale inhibitor is recommended.

Steamers

Steam generating equipment is very problematic for food service establishments. Since the water is turned to steam, leaving behind the dissolved solids, scaling is chronic with steamers. Unable to solve the problem with scale inhibitors, carbon filtration, mechanical filtration or softening, the food service industry has turned to commercial reverse osmosis systems for remediation.

The most commonly used ROs are 90 to 400 gallon-per-day systems. Food service equipment companies generally do not have the technical and application skills necessary for commercial RO technology, although they can easily use carbon filtration. Unaware of the benefits of RO, food service enterprises can spend thousands of dollars for unneeded cleanings and boiler equipment replacements.

The sizing of an RO system is critical. It is important to know the output of the steam generator, which translates into water use. All other factors that affect RO performance have to be considered, especially feed temperature, TDS, operating pressure and membrane production at standard conditions. The system is sized to exceed demand, and contain a 10 or 20 gallon hydropneumatic storage tank.

Since RO concentrates dissolved oxygen (DO), and the pH is slightly acidic, it is critical to use post-treatment on an RO system to reduce the water's aggressiveness. Cartridges with calcite, silicate/phosphate blends and/or phosphates are used for post-treatment. The choice depends on many factors, including DO content, pH and the steamer's construction.

Using a central systems approach, one RO system can feed steamers, post-mix machines, coffee service, ice machines and/or ingredient water (used for preparing foods.) In such cases, sizing is critical where peak demand and make-up have to be carefully evaluated. RO systems generally range from 1,500 to 2,800 gpd, with up to 500 gallon storage, and will include float activation of the RO, and a repressurization system.

Although the RO system is centrally located, post treatment varies and is usually located at the point-of-use. A central system may be the ultimate application of water quality improvement to food service applications, but it is uncommon because of the lack of qualified personnel.[11]

Water Vending Machines

Connected to an approved public drinking water supply, water vending machines process water using typical POU/POE technology. The types of unit processes used in the machines may vary, but ultraviolet or ozone treatment should always be included to guard against microbiological contamination.   In fact, most state regulations mandate some type of disinfection.

Typical purifier options include carbon, microfiltration, reverse osmosis (RO) and deionization (DI).   Depending on water processor selection, the machine will deliver a variety of product waters. The most common products are:

- Drinking water with less than 500 parts per million (ppm) total dissolved solids (TDS).

- Purified drinking water which is usually RO processed.

- Deionized water or purified water - for distilled water purposes with less than 10 ppm TDS, usually by DI.[12]

Some machines can deliver two of these products, offering the consumer a choice.

Because the devices are purifying water for human consumption, state or local health departments may require annual licensing and periodic testing of the equipment.   Individual machine licenses are usually easy and inexpensive to obtain, providing the vending machine design has been third-party validated for water processor effectiveness and ability to meet standards.

Private organizations such as the National Automatic Merchandising Association (NAMA) offer equipment validation test programs.   In most cases, it is the manufacturer who submits the equipment to NAMA in order to obtain the required listing.   When shopping for water vending equipment, one should look for the NAMA or a similar listing.[13]

Well-designed machines require very little technical service: an occasional filter change, an annual UV lamp replacement, and new RO membranes every 2-3 years.   Machines equipped with DI treatment systems require a periodic resin tank exchange for regeneration as well.   These procedures are all handled in the field by service personnel.

A vending machine of rugged steel construction will last many years. Every 4-5 years, it may require cosmetic refurbishment.   Figure 28 is a schematic of a water vending machine and the use of several unit processes.   Figure 29 shows a vending machine located at a municipal water treatment plant.

Some machines are controlled electro-mechanically while others use the newer microprocessor design.   Local water quality and availability will dictate design of a water vending machine.   High traffic locations such as supermarkets, large drug stores, and variety stores are typical locations where water vending machines can be found.

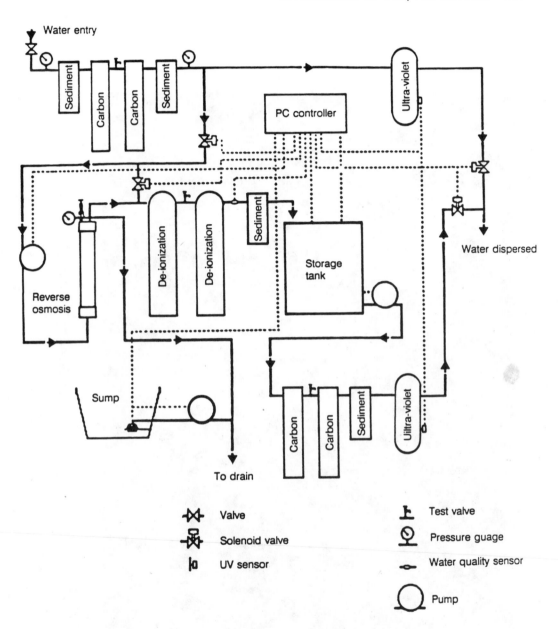

Figure 28. Water Vending Machine Schematic

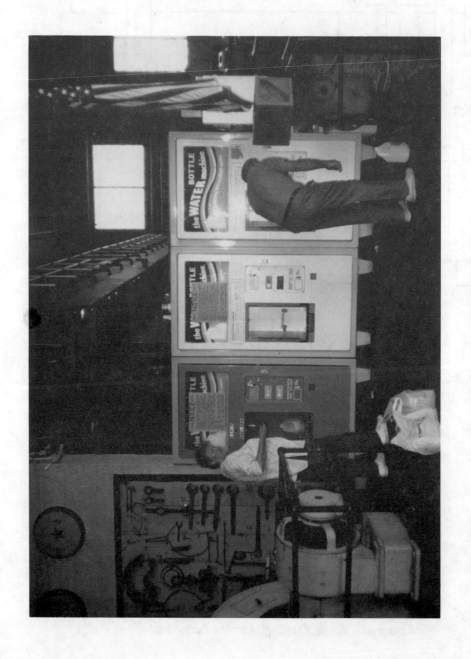

Figure 29.   Water Vending Machine

IEX systems in vending machines are less expensive than RO with less repairs. Tank regeneration may be necessary every few weeks in high TDS areas but can be cost effective in 600-700 gallons per day capacity machines.

Reverse osmosis involves a different kind of service over time than that of IEX. RO systems are physically concise and one with a 600-800 gallon per day capacity is ideally suited to a vending application. However, the slower RO flow rate requires a storage tank and a delivery pump. The RO pump is activated by a float in the storage tank.

Size of the storage tank is an important consideration because if the tank becomes depleted, the machine will sell out until it is replenished. In many areas where IEX resin is either unavailable or too expensive, RO is the system of choice.

Combination RO/IEX systems offer the best of both worlds. As you can see in Figure 28, reverse osmosis is often used as a DI pre-filter to increase the life of the IEX tanks. IEX helps to further reduce TDS from the RO to below the 10 parts per million NAMA standard for purified water. System configurations can be adapted to suit the water supply. For example, in especially high TDS areas, RO water along with sediment filters, carbon and ultraviolet can be used for drinking water. IEX can be added for purified water. If TDS is below 400, sediment filters, carbon and UV alone will usually provide a fine-tasting drinking water.[13]

Bottled Water Applications

Because of the dramatic increase in bottled water sales, new bottling companies as well as renovation of existing plants will also increase. Most bottling plants are or will be utilizing ozone as the primary oxidant along with most of the unit processes found together in water vending machines except at a larger scale.[14]   Ozone may be used first in the treatment scheme or following media filtration depending on the particular source quality and finished water quality goals. Filtration media selection depends on types and levels of contaminants to be treated, flow rates, capacity, and cleaning requirements. Granular filtration could include:

- Silica sand (mud, sand, rust, grit removal)

- Aluminum Silicate (mud, sand, rust, grit removal)

- Multi-media (ozone destruction, fine particles)

- GAC (organic compounds)

Membrane filtration can be used for solid/liquid separation. Microfiltration, ultrafiltration, and reverse osmosis filtration are able to remove successively smaller particles and are well-suited for post-media filtration. RO can produce very low TDS water as well as remove some SOCs. Water being sold as mineral, natural, or spring water would not include RO in the treatment process. RO does appear to be a cost effective alternative in bottled water applications.

Car Wash Systems

Spot-free rinsing is now frequently incorporated in car and truck washing systems. The use of RO water is superior to softened water and more economical then deionizing systems.[16] As water becomes more scarce in some areas and water discharge standards become more stringent, concentrate from RO systems can become a problem.

A sewer tie-in fee for a new car wash can cost more than $100,000. Many sites desirable for a car wash do not allow discharge of any water, including RO concentrate, so ways to reuse it are needed.

Reclamation systems, in conjunction with RO concentrate reuse, would enable potential car wash owners to locate their businesses on land unavailable to enterprises that discharge water. The average recovery rate of some RO systems is 25 percent, with 75 percent of the water wasted.

Concentrate Uses

One way to conserve water is to place several small membranes in a series instead of using one or two large membranes. This can increase the recovery rate to as high as 75 percent, which means only 25 percent becomes concentrate (water sent to the drain or reused).

This concentrate can be used in car washes as the initial rinse to remove soap from the vehicle. By storing the concentrate water in a tank and using it during the initial rinse cycle, all the water from the RO is used.

RO concentrate, although high in total dissolved solids, removes soap from vehicles because it is soft water. Spotting, which usually occurs only when the TDS is more than 70 parts per million (ppm), is not a concern. Most RO systems remove 98 percent of TDS.

Water with TDS as high as 5,000 ppm can be made spot-free by passing through the RO membrane. When a vehicle is rinsed with about five gallons of RO water, for example, the high TDS water from the initial concentrate rinse is not a factor.

A blending system in the logic control system can increase the recovery rate. The TDS is measured with a monitor, and blended with softened feed water to raise the TDS to a level slightly lower than 70 ppm. Use PVC pipe for low TDS water because metal piping will corrode with this "aggressive" water.

Water can be conserved by increasing the recovery rate, reusing the concentrate water and blending water to obtain a specific TDS. When treating a water supply with 40 ppm TDS, for example, the RO can be expected to deliver water at about 8 ppm, assuming 98 percent removal. The TDS can be raised to about 60 ppm by blending this water with softened water at 400 ppm to yield spot-free water and reduce the amount of makeup water required. Hence, the system's recovery rate can be increased.

Markets and applications change daily in the water treatment industry, and car wash and other business owners must stay abreast of these regulations to maintain a position in their area.

There is no consistency to federal, state, or local discharge regulations. Should new standards be promulgated by the USEPA, it is difficult to say how stringent the standards will be or when they will be enforced, or how strictly different states will enforce those standards.[17]

Horticultural Nurseries

In many geographical areas, tree and plant nurseries already have some alternate source of water low in mineral salts -- whether it be supplied in bulk or through the use of ion exchange resin tanks. RO is becoming increasingly popular as the preferred method because of its low operating expense.

Studies show that most seedlings and indoor potted plants grow healthier, and cut flowers last significantly longer in water that is low in mineral salts and free of chlorine.

Medical Applications

Research and medical laboratories, pharmaceutical, and cosmetic companies have been producing very high quality water for years. Their specific applications include:

- kidney dialysis
- laboratory diagnostics
- medical device rinsing
- pharmaceutical rinsing
- pharmaceutical production
- pharmaceutical product cooling
- cosmetics manufacturing

Kidney Dialysis. RO systems are already being used extensively for making purified water that is used throughout the dialysis procedure. There are two areas in dialysis to address with RO systems. One is the large central hospital dialysis system that requires sophisticated engineering and a regular detailed competent service and testing program. The other area is in home dialysis which is a growing trend. Each patient in many cases will require a softener as pretreatment. Various medical programs, such as Medicare, usually fund the purchase of such equipment.

Medical Laboratories. Most research and medical labs require a source of deionized water of 1 Meg Ohm or better. These needs are usually met with exchange deionizer tanks and less often with distillers.

Using an RO system in conjunction with resin bed deionizers will not only provide a better quality of water (especially reduced levels of particulate matter, bacteria and organics) but can substantially reduce the cost per gallon of this high purity water depending on the quantity of water used and the ion content of the raw water. In the near future, many laboratory and pharmaceutical applications may utilize special RO systems eliminating the need for ion exchange resins.[11]

Industrial Applications

Industrial applications may be considered more as pre-treatment for chemical or production processing. These include:

- electronic (semiconductor rinsing)
- printed circuit rinsing
- metal finishing
- boiler feed
- cooling tower feed

An example of use of drinking water treatment technologies for industrial application is the AT&T microelectronics plant in Orlando, Florida. The 400,000 gpd (18-L/s) plant produces ultrapure water for the electronics industry by using a variety of treatment processes, including reverse osmosis and ion exchange.

A pretreatment system removes foulants from the ground-water source. The system consists of a packaged coagulation-filtration system and a microfilter. Effluent is then treated by a three-stage high-pressure reverse-osmosis unit using spiral-wound, cellulose triacetate membranes. To control scaling, acid and an antiscalent are added prior to the RO unit. The product water from the RO unit is further treated by a mixed-bed cation-anion exchange system. The resulting ultrapure water is then degassed and treated with ultraviolet sterilization.[18]

Metal finishing plants such as metal platers and anodizers require high purity water also in their processes. RO systems are now being used to replace or in conjunction with traditional ion exchange.

Table 15 displays several of the applications briefly described in this chapter and provides further information on their use and constraints.

Farm Applications

There is a great need for water treatment in farming -- agricultural areas. With increased use of agricultural chemicals, wells can become contaminated with nitrates, arsenic, mineral salts, pesticides and herbicides. These impurities must be reduced to safe levels for both household use and the feeding of most young livestock. These individual wells may need POU/POE devices to produce safe drinking water. For individual rural homeowners without the benefit of a central treatment system, POU/POE may be the only viable solution for producing potable water.

TABLE 15.   COMMERCIAL AND LIGHT INDUSTRIAL APPLICATIONS OF POU/POE[19]

USE	QUALITY REQUIREMENTS	QUALITY STANDARDS	QUANTITY REQUIREMENTS (TYPICAL)	COMPETITIVE PROCESSES
CONSUMER/COMMERCIAL				
Brackish water-Potable	500 ppm TDS	USPHS, WHO	1 gpd to 5,000,000 gpd	DI Distillation, ED
Seawater Desalting-Potable	500 ppm TDS	USPHS, WHO	1 gpd to 5,000,000 gpd	DI Distillation
Bottled Water	500 ppm TDS	USPHS, WHO	20 gpm to 100 gpm	DI Distillation
Ice Manufacturing	50 ppm TDS	USPHS, WHO	1 gpm to 20 gpm	Softening, DI
Car Wash Rinsing	100 ppm TDS		½ gpm to 2 gpm	Softening, DI
MEDICAL				
Kidney dialysis	50 ppm TDS, bacteria, virus free	AAMI (Proposed)	10 gph to 10 gpm	Softening, DI, Membrane-Filtration
Laboratory	0.01 ppm TDS, bacteria, virus free	ASTM	10 gph to 10 gpm	DI
Pharmaceutical	10 ppm TDS, bacteria virus free	USP	1 gpm to 50 gpm	DI Distillation
Cosmetics	10 ppm TDS, bacteria, virus free	USP	1 gpm to 50 gpm	DI Distillation
INDUSTRIAL				
Boiler Feed	Depends on Boiler pressure		½ gpm to 30 gpm	Softening, DI
Humidification	10 ppm TDS		½ gpm to 30 gpm	Softening, DI
Rinsing	Depends on application	Depends on application	½ gpm to 100 gpm	Softening, DI
Chemicals Production	10 ppm TDS Low organics		½ gpm to 50 gpm	DI, Distillation

Abbreviations:

TDS = Total dissolved solids, AAMI = Association for the Advancement of Medical Instrumentation, ASTM = Amer Society for Testing and Materials, USP = United States Pharmacopoeia, USPHS = Unites States Public Health Servi WHO = World Health Organization, gpm = gallons per minute, gph = gallons per hour, gpd = gallons per day, deionization, ED = Electrodialysis

REFERENCES

1. Lykins, Jr., B. W. and Baier, J. H., "Point-of-Use/Point-of-Entry Systems for Removing Volatile Organic Compounds from Drinking Water", Chapter 16, Significance and Treatment of Volatile Organic Compounds in Water Supplies, Lewis Publishers, Inc. (1990).

2. Chambers, C. D. and Janszen, T. A., "Point-of-Entry Drinking Water Treatment Systems For Superfund Applications", Risk Reduction Engineering Laboratory , Cincinnati, OH, Project Officer: Mary K. Stinson, EPA/600/S2-89/027, (February, 1990).

3. Gentry, J. K. and Dean, W. G., "Removal of EDB From Domestic Water Supplies With GAC Filtration Systems", Bureau of Operations, FLorida Department of Environmental Regulation, Tallahassee, Florida, (1984).

4. Lowry, J. D., "Aeration vs. GAC For Radon Removal", Water Technology, 13(4):32-37 (1990).

5. Lee, R., "Experience With Cartridge Filters at Longmire, Fiscal Year 1983 and 1984", National Park Service, Pacific Northwest Region, Seattle, Washington, (September, 1984).

6. Sorber, C. A., Malina, J. F., Jr., and Sagik, B. P., "Virus Rejection by the Reverse Osmosis - Ultrafiltration Processes", Water Research, 6:1377-1388 (1972).

7. Valcik, J. A., Phull, K. K., and Miller, R D., "Use of POE/POU Technology at U.S. Army Installations", Proceedings of ASCE National Conference on Environmental Engineering, Austin, Texas, July 10-12, 1989.

8. Hinterberger, J. A., Lindsten, D. C., and Ford, A., "Use of Reverse Osmosis and Ultrafiltration For Removing Microorganisms From Water", Army Mobility Equipment Research and Development Center, Fort Belvoir, Virginia, (September 1974).

9. Cole, R. W., "Effectiveness Of Commercially Available Home Water Purification Systems For Removing Organic Contaminants", Engineering & Services Laboratory, Air Force Engineering & Services Center, Tyndall Air Force Base, Florida, (May 1986).

10. Slovak, J. and Slovak, R., "Tapping the Reverse Osmosis Market: A Roadmap to Success", Water Conditioning, (February, 1983).

11. Montemarano, J., "Commercial Food Application", Water Technology, pp. 42-49, (September, 1990).

12. Barcus, H. C., "Expand Your Business with Water Vending Machines", Water Conditioning & Purification, 32(11):34-38, (December, 1990).

13. Barbaccia, L., "Shopping for Features in Water Vending Machines", *Water Conditioning & Purification*, 32(11):52-58, (December, 1990).

14. Andrews, S., "Standards of Applications for Bottling Plant Ozonated Water", *Water Conditioning & Purification*, 32(10):38-44, (October, 1990).

15. Andrews, S., "Standards of Applications for Bottling Plant Ozonated Water, Part II: Filtration Selection", *Water Conditioning & Purification*, 32(10):61-66, (November, 1990).

16. Slovak, J. and Slovak, R., "Tapping the Reverse Osmosis Market: The Commercial/Light Industrial Market", *Water Conditioning & Purification*, p. 19, (June, 1983).

17. Karambis, G. T., "Reclamation Systems for Commercial ROs", *Water Technology*, 13(4):30-31, (April, 1990).

18. *American Water Works Association Mainstream*, p. 7, (December, 1990).

19. Cartwright, P. S., "Demystifying RO Water Treatment", *Water Technology*, 5(5), (August, 1982).

CHAPTER 4

INTRODUCTION

With any water treatment system, there are certain operational and maintenance considerations that one has to be aware of to efficiently provide safe drinking water. Point-of-use and point-of-entry systems are no exception. All too often, the homeowner has the impression that they have a treatment device for their drinking water and therefore they are protected from any harmful contaminants. Point-of-use and point-of-entry devices can not be installed and forgotten. Proper operation, maintenance, and monitoring of point-of-use/point-of-entry systems is required in order for them to provide adequate performance. A major component of any POU/POE operation is monitoring. Many of the chemicals that are removed are odorless, colorless and tasteless -- but still toxic or carcinogenic.

INSTALLATION

Various types of installers may be used depending on the homeowner preference: equipment dealer or subcontracted local plumber. Either must have proper training to install the device. A treatment unit may require a variety of installation skills, including:

- electric -- solenoid valves, power generation (RO)
- plumbing -- soldering, waste disposal (cross-connection)
- mechanical -- valves, pipe layout, selection of special regulators or cross-connection devices
- structural -- location and physical installation

It has been recommended that installers be licensed from local or state government.[1] The requirements for that license would include satisfactory training from a reputable manufacturer, training in cross-connection and backflow prevention, and proven demonstration of unit installation.

Long Island investigators found positive coliforms in effluents after installation (resamples were negative). They recommended that manufacturers develop an adequate disinfection procedure before shipping devices

147

and that all parts and units be disinfected at the time of installation at the site.[2] The use of backwashing devices is not recommended for GAC units and should there be a concern about particle buildup on the carbon bed, a prefilter should be installed.

Therefore, one should consider the following factors in selecting an installation contractor.[2]

  • Demonstrated experience in installing point-of-use/point-of-entry treatment devices.
  • Conformance with applicable plumbing codes.
  • Liability for property damage during installation.
  • Accessibility for service calls.
  • Contractor's responsibility for minor adjustments after installation.
  • Price quote basis (hourly rate versus per unit rate).
  • Documentation of installations and provide references.

OPERATION

Activated Carbon

Although the adsorption characteristics of individual organics vary, they are amenable to removal by activated carbon adsorption for a period of time or gallons of water treated. The amount of water that can be treated depends on the influent contaminant concentration, the amount of carbon in the treatment unit, and the time the contaminant is in contact with the carbon (contact time). The removal efficiency of the carbon system will deteriorate with use, and eventually breakthrough will occur. The treatment units usually specify a rated *gallons treated* that most manufacturers stress should not be exceeded. If exceeded, the potential exists for carbon exhaustion and desorption of organics in higher concentrations than those in the untreated water to the unit. Figure 1 demonstrates the variability of contaminant breakthrough based on different contact times with an average trichloroethylene (TCE) influent of 75 μg/L.[3] Small variations in hydraulic loadings can result in very different levels and duration of exposure to contamination.

Figure 2 displays how competition for adsorption sites leads to differing contaminant breakthroughs. Tetrachloroethylene (PCE) is much more strongly adsorbed to carbon than cis 1,2-dichloroethylene (cis 1,2,-DCE). Thus, PCE is less likely to breakthrough prematurely or in response to competition from other contaminants in a multi-contaminated source water.

Exhaustion of the GAC bed can yield varying levels of contaminant removal. Figures 3 and 4 show a bench-scale study done by a POU manufacturer that illustrates how performance can vary depending on the service life of the treatment unit.[4]

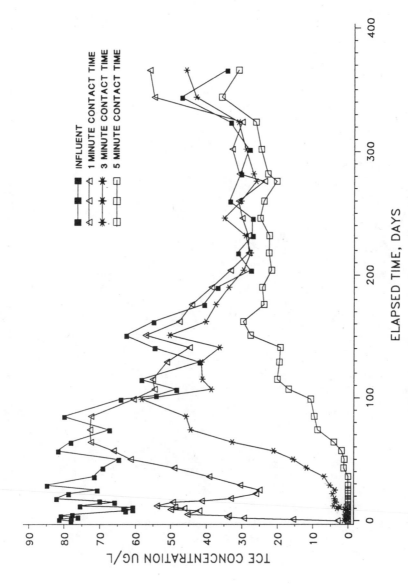

FIGURE 1. EFFECT OF VARIATIONS IN CARBON CONTACT TIME

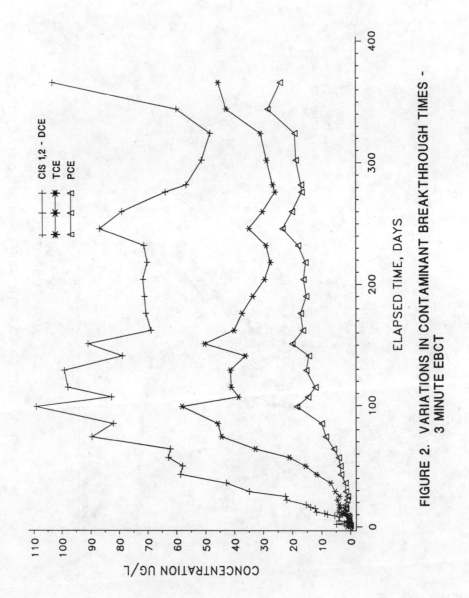

FIGURE 2. VARIATIONS IN CONTAMINANT BREAKTHROUGH TIMES - 3 MINUTE EBCT

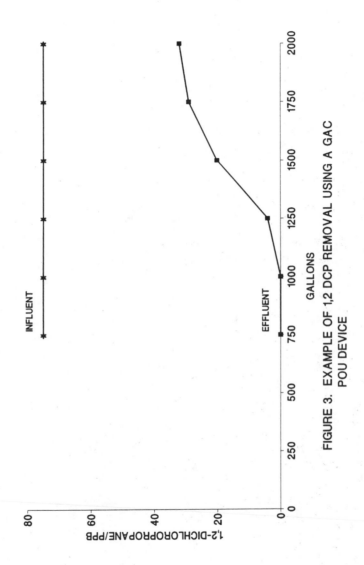

FIGURE 3. EXAMPLE OF 1,2 DCP REMOVAL USING A GAC POU DEVICE

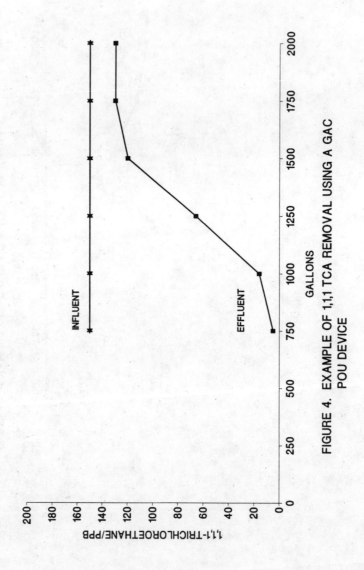

FIGURE 4. EXAMPLE OF 1,1,1 TCA REMOVAL USING A GAC
POU DEVICE

Operating for Radon Removal

Exposure to gamma radiation, a decay product of radon-222, must be considered when utilizing GAC treatment technology. Two general relationships have been derived regarding the amount of gamma radiation and the design and placement of the carbon tanks. The first is the maximum gamma exposure rate occurs on the outer shell of the GAC vessel as a function of the influent radon (Rn) activity. The mathematical relationship is nearly linear and is expressed as:

Max. Gamma = Rn influent / 10,360

where: Rn influent = the raw water Rn concentration in pCi/L
          Max. Gamma = mR/hr.

The maximum gamma rate was found very close to the top of the carbon bed on the outer surface of the vessel. The second relationship is the horizontal distance required to meet certain gamma exposure rates as a function of the raw water Rn activity. Federal guidelines recommend no more than 170 mR/yr (residential) and 500 mR/yr (public) over background levels. Lowry et al. (1988) suggests the following guidelines in a household:[5]

1.   In a non-living area, such as a cellar, the technical need for shielding becomes a concern at a Rn level of 100,000 pCi/L. To minimize any exposure, a shield should be used above a level of Rn of 20,000 pCi/L.

2.   In a living area, there is a technical need for shielding for GAC units treating Rn levels of greater than 20,000 pCi/L.

3.   Beyond the technical requirements, it may be a good general practice to shield GAC units treating wells containing Rn at greater than approximately 5,000 pCi/L.

The homeowner is oftentimes responsible for determining the criteria and risk associated with these units. This situation could lead to inappropriate installations of POE GAC for high Rn water supplies.

Shielding has been done on a number of POE/GAC units in the field. Lead, water, and other materials have been used. However, water represents the most cost effective solution in many cases.

The data in Figure 5 can be compared to the unshielded data shown in Figure 6. A comparison of the shielded and unshielded cases illustrates the effectiveness of the water shield. This shield was comprised of a simple 24 inch diameter polyethylene brine tank containing the GAC vessel and water at a level equal to the top of the GAC vessel.

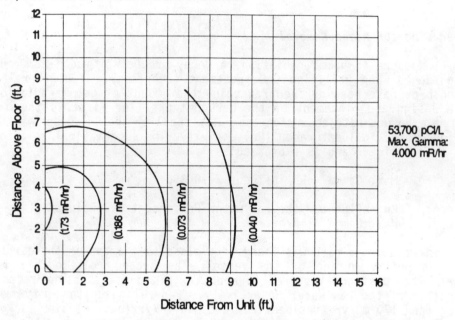

Figure 5. Gamma Exposure Rate Field at Site Without Water Shield

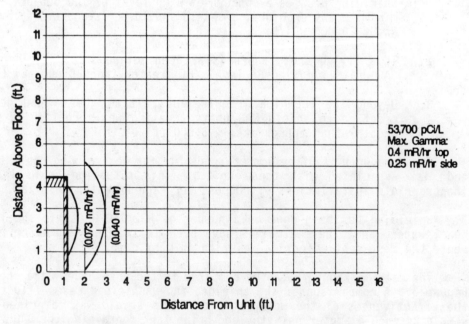

Figure 6. Reduced Gamma Exposure Due to 24" Water Shield

At this site, the GAC vessel was positioned inside the water tank to one specific side closest to the structure's foundation wall, to provide the maximum amount of water shield in the horizontal direction out into the room.   The details of suggested shielding guidelines are given in Figure 7.   Figure 8 shows what a typical installation might look like.

As mentioned previously, carbon will become exhausted and therefore replacement will become necessary.   This creates another concern; proper disposal of the exhausted or spent carbon.   One might think that this is not a problem considering the small amounts of carbon involved but consider the following scenario presented by Montemarano.[6]   The amount of carbon expended from a point-of-use device is on the order of a few pounds a year.   On the other hand, a point-of-entry system with a 1.5 cu. ft. tank has more than 50 lbs. of expended carbon per change even with modest dewatering.   At an average of two changes per year for VOCs, this amounts to 100 lbs. of carbon for each 1.5 cu. ft. tank.   Consider a small community of 500 homes that chooses to use point-of-entry treatment as an interim treatment method to produce potable water from an organically contaminated source water until an alternative water source is developed which is on the average of two to five years.   Someone, whether the water treatment contractor, local or State health department, etc., is faced with the problem of disposing of 25 tons of exhausted carbon that is saturated with organic chemical contaminants.   Disposal methods for exhausted carbon include landfilling, reactivation, and incineration.

Many communities and some States are restricting landfilling of exhausted carbon unless certain criteria are met.   On Long Island, New York, carbon is required to be disposed of in an environmentally safe manner, landfilling is not permitted.   Spent carbon is shipped out of state to the manufacturer where it is regenerated for reuse in industry but not for potable use.[7]

Membranes

As discussed previously, membranes can be a viable treatment option for the homeowner.   However, there are certain operational conditions that have to be considered.   In order to protect the membranes from fouling, adequate pretreatment using a high quality micron filter is required. Deposits of colloidal material, oils or precipitated salts will cause a flux decline.   Pretreatment using a 5 μm filter is often used to reduce membrane fouling.   Most reverse osmosis membranes have a life of 3 to 5 years if they are not fouled or damaged.   Reverse osmosis systems will not tolerate high concentrations of iron (>0.1 mg/L), manganese (>0.5 mg/L), hydrogen sulfide (>0 mg/L), hardness (>20 gpg in most cases), and turbidity (>1 NTU).[8]   Another important operational factor to be aware of when using membranes is the water or feed pressure.   This pressure should be between 40 and 100 psig to force water through the membrane.[9]   While this may not be a problem in most homes, occasionally additional pressure has to be supplied to the membrane system.   Rejection of contaminants depends on the pressure and generally, the higher the pressure the higher

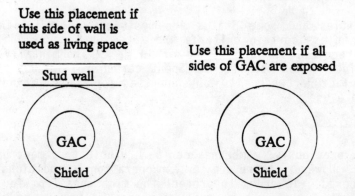

Figure 7.   Placement Alternatives of GAC in Shield

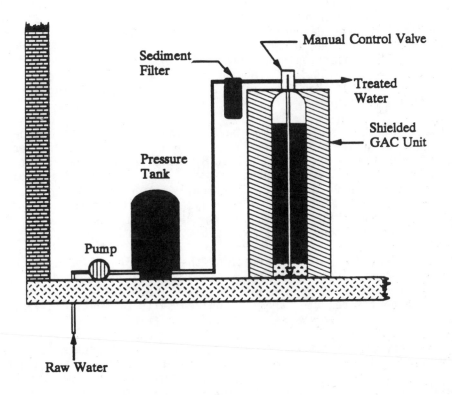

Figure 8.   Typical GAC Installation for Radon Removal

the percent rejection. If the total dissolved solids in the influent or source water increases more water pressure is required to maintain proper operation of the membrane. For example, every 100 mg/L (ppm) of total dissolved solids requires about one psig just to overcome osmotic pressure.[8] Also, the flux rate is directly proportional to the water pressure. Another factor that can effect the recovery rate is flow control.

One of the most commonly overlooked factors in using membranes is the temperature of the source or feed water. Most reverse osmosis membranes have claimed production rates in gallons-per-day determined at the industry standard of 77°F.[8,10] Therefore, it is important to know that membrane water production rate increases or decreases 1.5% to 2% per °F increase or decrease, depending on the type of membrane. A 50°F water supply, which may not be uncommon, can result in as much as a 54% decrease in rate of water production from what the membrane manufacturer indicated (2% X (77°F - 50°F) = 54%).[10] Another limiting factor may be materials of construction other than the membrane that can fail under high pressure at elevated temperature.

The pH of the feed water may have a significant effect on the durability and performance of the membrane. For cellulose acetate type membranes, the pH limits are normally 3 to 8. At this range the rate of membrane hydrolysis can be tolerated.[10,11,12] The membrane has a minimum hydrolyze rate at a pH of about 5. Basically, cellulosic based membranes usually will slowly deteriorate or hydrolyze at pH levels above 8 and will last longer in the slightly acid pH range.[10] For polyamide hollow fiber membranes, the recommended pH limits are 4 to 11.

Feed water pH has little effect on the rejection of salts of strong acids and strong bases. However, it may have a pronounced effect on rejection of relatively weak acids and bases. Generally, the nonionized species are poorly rejected and the ionized salts of weak acids and weak bases are well-rejected.[11]

Most reverse osmosis systems contain a post-treatment storage reservoir to collect treated water prior to use. These tanks are often pressurized and can result in a back pressure on the membrane if not properly operated. If this occurs, the pressure drop across the membrane will decrease with a correspondingly decrease in contaminant rejection. Another problem that may occur in the storage reservoir unless adequate disinfection is maintained is bacteria growth. If this occurs, periodic cleaning may be required. Also, once the storage reservoir is filled in some systems, surplus water is discarded resulting in excessive water loss from the system.

The positioning of the post-carbon filter after the storage tank, common to all RO/GAC systems, removes taste/odors from the tank and plastic tubing just prior to dispensing. The GAC filter can also act as a second barrier for chemical contaminants should the membrane have a flaw. The water travels through the carbon filter at the rate at which it is being dispensed, approximately 0.5 gpm.

In cases involving high levels of organic chemical contaminants like VOCs, choosing the proper degree of carbon filtration is critical. An RO/GAC system with a longer contact time removes organic chemical compounds not removed by a semipermeable membrane or the standard post-carbon filter.[8]

As usual, disinfection should be applied to insure microbiological water quality following the carbon filter.

One of the potential major disadvantages of reverse osmosis treatment is the disposal of the concentrate or reject water. Since most reverse osmosis units operate with a 10% to 30% recovery rate, and a continuous water flow is needed to maintain osmotic potential, up to 9 gallons of concentrate can flow to the drain for each gallon collected for consumption.[9] This may not appear to be a concern but if a household has a septic tank that is stressed to near capacity or leach lines are partially plugged, this could be a serious problem.

Ion Exchange

In order to obtain satisfactory operation of an ion exchange unit, one of the first challenges is to select the appropriate resin. This selection process depends on several factors but a basic requirement is a knowledge of the composition of water to be treated and the effluent requirement. For a cation resin, the performance can be affected by chlorine or other disinfectants used to control bacterial growth in water systems.[13] These oxidizing agents degrade the resin beads crosslinkage causing an increase in water content and excessive swelling. Also, the beads may become mushy and break into fine particles thus reducing contaminant removal. This will result in increased pressure loss across the softener bed and result in poor regeneration. Using resins of increased crosslinkage have been reported to extend a resin's life by two to three fold.[13]

Other potential problems in water softeners are suspended solids, manganese, and iron. Control of suspended solids may require prefiltration. Iron and manganese tend to enter the resin bed as soluble salts that oxidize, causing a sticky deposit which has to be removed for the beds to remain functional. These oxides can be removed by a high backwash flow. Water temperature is also an important factor in selecting an ion exchange resin. Resin bed expansion will increase significantly during backwash as the water temperature decreases. High density resins have been reported to offset some of this effect.[13]

One of the major operational concerns with ion exchange is the need to dispose of a concentrated brine which can often have high levels of sodium chloride. Dilution in the septic system may limit some of the treatment's adverse effects. Other concerns are that the product water from a softener may become corrosive to household plumbing and result in elevated levels of copper, iron, or lead in addition to shortening the life of the plumbing fixtures.[9] Another operational concern is the possibility of selective contaminant removal. With most anion exchange resins, sulfate is preferred over nitrate which reduces the nitrate exchange capacity of the resin. The health significance of this is that with high sulfate concentrations previously exchanged, nitrate ions can be released back into the product or effluent water.

Hydraulic loading rates, bed volume, particle size, and pressure requirements are all critical in maximizing removal. Optimal water quality parameters are listed in Table 1.

TABLE 1.   OPTIMAL INFLUENT WATER QUALITY PARAMETERS
FOR EFFICIENT ION EXCHANGE OPERATION[9]

Parameter	Characteristic
Iron	< 5 mg/L
pH	Contaminant specific
Turbidity	< 5 mg/L
Temperature	< 140°F

Anion exchange resins can be subject to fouling from the lead cation exchange resin. Organic fouling and degradation products from oxidation attack on the cation structure can also cause fouling.[11]

Distillation

With distillation, few operational problems are likely to be encountered. Improperly designed distillers may have potential for shock or fire hazards because of high operating temperatures and electrical requirements. This can be avoided by ensuring that the unit has proper electrical safety approvals including Underwriters Laboratory (U.L.) approval. Another hazard can come from touching heated parts. Unevaporated contaminants remaining in the boiling chamber have to be flushed-out regularly to a septic or sewer system. Even with regular removal of the residual water containing various contaminants, a scale of calcium and magnesium will collect at the bottom of the boiling chamber.[14] This scale, which will also collect on the heating elements, is likely to be difficult to remove requiring hand scrubbing or application of an acid. Cleaning frequency will depend on the hardness and total dissolved solids of the feed water. Although the distilled or purified water container is relatively free of contaminants, after a period of time it will become microbially contaminated.   This mainly occurs because the homeowner neglects routine washing and cleaning of the container because it only contains purified water.   However, they fail to realize that these containers are not sterilized.   Optimal water quality parameters are listed in Table 2.

TABLE 2.   OPTIMAL INFLUENT WATER QUALITY PARAMETERS
FOR EFFICIENT DISTILLATION OPERATION[9]

Parameter	Characteristic
Hardness	< 200 mg/L as $CaCO_3$
Temperature	< 70°F

Air Stripping

As with distillation, few problems are likely to be encountered with packed tower air stripping as long as the system has been designed properly.   The correct air-to-water ratio is necessary.   Otherwise too much air may blow water out of the top of the system and not enough air will result in poor removal of volatile organics.

There are various types of packing material available for air strippers as shown in Figure 9.   Therefore, one should determine which material is best for each situation.   Also, if dissolved iron and manganese is in the water it will precipitate on the packing material and periodic cleaning with acid may be required or pretreatment can be installed to remove the iron and manganese before aeration.   Diffused bubble aerators need to be concerned with clogging of the air holes.

Critical operating parameters are flow rate and pressure drop. Liquid channeling down the tower wall is another concern that can reduce removal efficiency.   Aeration can produce a more aggressive water that could cause or increase the rate of corrosion of water heaters, plumbing, and joint materials, thus subjecting consumers to those concerns.   The optimal water quality parameters are listed in Table 3.   Clogging of packing material, colored water, or staining of water fixtures in the home can result from high levels of those water quality parameters listed below.

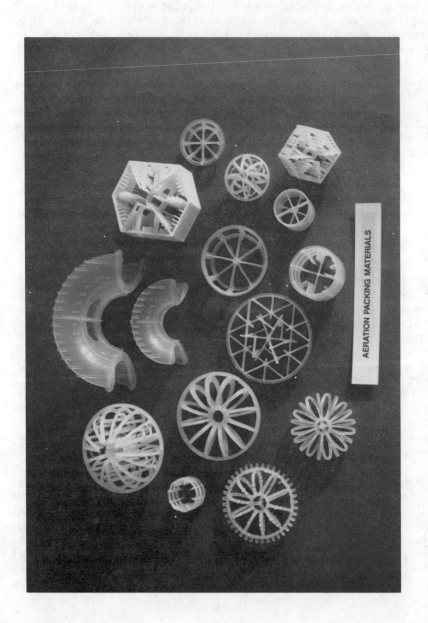

Figure 9.   Different Types of Packing Material for Air Strippers

TABLE 3.   OPTIMAL INFLUENT WATER QUALITY PARAMETERS
FOR EFFICIENT OPERATION OF AIR STRIPPERS[9]

Parameter	Characteristic
Hardness	Site and design specific
Suspended Solids	Site and design specific
Iron	< Secondary MCL
Manganese	< Secondary MCL
Temperature	Lower temperatures decrease removal

Activated Alumina

One of the major operational considerations when using activated alumina is pH. The activated alumina system should be operated between a pH range of 5 to 6 for optimum efficiency. Problems can occur if the activated alumina is used and the media fines are wetted but the filter is subsequently not used for an extended time period. This will cause the media to cement. It has been reported that noncoliform bacteria growth on the activated alumina can also be a problem when the influent or feed water contains little or no chlorine residual.[9] Coliform bacteria do not usually increase through the media and may even decrease. When elevated levels or noncoliform bacteria occur from the activated alumina bed, flushing with significant quantities of bacteria free water appears to reduce these levels.

Operating the unit at higher than recommended flow rates will reduce contact time, thus reducing removal efficiency and/or adsorption capacity. Highly variable influent quality will also result in sooner than expected breakthroughs. Other pertinent water quality parameters include alkalinity, pH, sulfate, iron, turbidity, TDS, suspended solids, and temperature. Failure to provide source water within the ranges presented below in Table 4 could result in less than optimal removal and exposure to the contaminant(s).

TABLE 4.   OPTIMAL INFLUENT WATER QUALITY PARAMETERS
FOR EFFICIENT OPERATION OF ACTIVATED ALUMINA SYSTEMS[9]

Parameter	Characteristic
Alkalinity	< 1,000 mg/L as $CaCO^3$
pH	< 6 for fluoride; = 7 for arsenic
Sulfate	< 250 mg/L
Iron	< MCL
Turbidity	< MCL
TDS	< MCL
Suspended Solids	< MCL
Temperature	1 - 30°C

Disinfection

Ultraviolet Light

The germicidal effect of an ultraviolet lamp on the light intensity and exposure time is shown by the following equation.[15]

$$N = N^0 \times Q^{-Et}$$

where:

N = number of surviving bacteria
N^0 = number of bacteria before treatment
E = ultraviolet light intensity
t = exposure time
Q = unit lethal exposure

The intensity or penetration of ultraviolet light through the water to be disinfected into the microbes is affected by several operational factors as shown below.

•   The optimum operating temperature of an ultraviolet lamp is 40°C or 104°F.   If the lamp is cooled by contact with the water, output may drop.   Therefore, a sleeve is usually used to separate the lamp from the water to maintain lamp operating temperature.

•   Ultraviolet light output is affected by flow rates, flow patterns in the reactor such as short circuiting, shadowing in the reactor, and lamp configuration in multiple lamp systems.[16]

•   A decrease in line voltage can cause a decrease in lamp output.

- Lamp ageing can also effect the light intensity. Newly in-stalled lamps can decrease in radiation by as much as 20% during the first 100 hours of operation but then normally maintain that level or maybe produce a slight change for about 7,500 hours of operation. After 7,500 hours, the lamp output is usually about 70% of that of a new lamp. Manufacturers normally recommend replacement of lamps after 7500 hours of service or every 10 to 12 months.

- The presence of turbidity, color, dissolved iron and organic compounds can also affect performance. Clumping of organisms or shielding of organisms from radiation by particulate matter will reduce the effectiveness of an ultraviolet device. Dis-solved iron could be deposited on the lamp which would interfere with the penetration of ultraviolet light into the water.

For ultraviolet to effectively disinfect drinking water, all glass parts must be clean. Prefiltration is required to minimize turbidity and provide maximum light penetration. Ultraviolet units should be placed after other types of treatment such as granular activated carbon, ion exchange softeners, or reverse osmosis to ensure that any organisms that may have grown in a filter or tank are exposed to ultraviolet light before going into the plumbing.

The Canadian government has suggested an upper limit of 1000 total coliforms/100 mL or 100 fecal coliforms/100 mL for use of UV treatment technology. Iron, calcium, and manganese can also interfere with UV penetration by depositing on the bulb or quartz sleeve and absorbing the light. Even levels below the MCLs can over time interfere with UV effi-ciency. Suspended solids and turbidity levels less than 10 mg/L and 5 NTU, respectively, have been recommended for most effective treatment. UV lamp output is strongly temperature dependent with optimum output at 40°C. Some type of flow regulator may be necessary to ensure that flow capacity of the unit is not exceeded, thus decreasing the water's exposure to the light and reducing efficiency and increasing exposure to microbio-logical contaminants.[15]

Ozone

In order to achieve adequate oxidation or disinfection with ozone, sufficient ozone dose and contact time is important. Because all waters to be treated are different, it is difficult to predict accurately the dosage of ozone required or the contact time needed for each application. However, this can be determined before ozone installation. For bacteria, virus, and cyst control, inactivation or destruction of these microor-ganisms can be determined by the Ct value. This is the product of dis-infection concentration (C in mg/L) times the contact time of the disin-fectant with the water (t in minutes). Water temperature and pH will effect the Ct values. The higher the water temperature, the lower the Ct value required to provide adequate disinfection. A summary of Ct values for inactivation of various microorganisms by ozone is shown in Table 5.[17]

TABLE 5.   SUMMARY OF Ct VALUES FOR INACTIVATION OF VARIOUS MICROORGANISMS BY OZONE

Microorganism	Ozone Ct Values pH (6-7)
E. coli	0.02
Polio 1	0.1 - 0.2
Rotavirus	0.006 - 0.06
Phage f^2	ND
G. lamblia cysts	0.5 - 0.6
G. muris cysts	1.8 - 2.0
Cryptosporidium parvum	5 - 10

All Ct values are for 99% inactivation at 5°C except for *Cryptosporidium parvum* which was for 99% inactivation at pH 7 and 25°C.

Most ozone generators use a high voltage electrical arc in the production of ozone and will therefore require electrical service. Typical electrical usage rates are in the range of 50 to 100 watts/h or about the same as a television set. Pretreatment filtration may be required prior to the ozone unit to remove turbidity or suspended solids which can interfere with ozone oxidation and disinfection. If the ozone unit is being used for inorganic oxidation, post-treatment filtration is often recommended to remove the oxidized metals such as iron and manganese. Also, because ozone gas can have adverse health effects, it is important to ensure that leaks from the generator system do not occur. In addition, any residual ozone in the vent gas should be destroyed or vented outside the home.

Other installation and operational concerns have been described by Andrews.[18] Some of these are listed below.

- To prevent possible ozone gas leakage into the home from point-of-entry systems, all ozone generators should have air drawn through under vacuum.

- To ensure proper air flow and detect potential down stream injection tubing cracks or breaks, all ozone generators should have air flow meters installed before the ozone generation chamber.

- All ozone generation chambers should be constructed of stainless steel.

- When used for disinfection, all ozone generators should be installed with air preparation equipment such as air dryers.

- All ozone generators should have corona arc or UV indicating lights.

- Consideration must be given to proper gas injection tubing, gaskets, valves, sealants, etc. because of the strong oxidizing power of ozone. The materials should consist mainly of teflon and stainless steel (316 or 304).

- Consideration must be given to proper gas injection tubing, gaskets, valves, sealants, etc. because of the strong oxidizing power of ozone. The materials should consist mainly of teflon and stainless steel (316 or 304).

- In-line check valves are needed to prevent water from backing-up into the unit through the injection tubing which can cause premature ozone generator failure. Sometimes, in order to provide more assurance against back-up, a manometer is used because it is difficult to find reliable check values.

- The piping material between the ozone injection point and the filter should have some chemical resistance qualities. Ozone injection tubing should consist of the following: teflon, 316 or 304 stainless steel, Kynar, ethylene-propylene terpolymer, polyvinylidene fluoride, Hypalon, Fluoropolymer, ceramic, glass, schedule 40 PVC, Viton, unplasticized PVC, and "Red polyurethane".

- For ease of operation and cost, ozone injectors should consist of venturi nozzles for most point-of-entry ozone applications provided that these nozzles are chemical resistant and provide efficient mass transfer.

- Ozone reactor tanks should be designed to: (1) provide adequate water volume within the tank to facilitate completion of desired chemical reactions, (2) provide for dispersion of unadsorbed ozone gas bubbles to enhance mass transfer efficiency, and (3) provide an efficient water flow pattern by using internal baffles in order to promote degassification of unabsorbed ozone gas.

Ozone contacting systems do not provide total (100%) transfer of ozone into the water being treated. Some ozone always passes through the ozone contactor (reactor). A well-designed ozonation system will provide for this eventuality by incorporating a gas collector and an ozone-destruction unit. Ozone can be destroyed thermally (heating to 300°C) or catalytically (by passing through GAC or manganese dioxide, for example). The treated air then can be discharged into the ambient atmosphere devoid of ozone.

Optimal operation conditions include proper ozone dosage, contact time, and ozone generation.   Important water quality considerations include a pH slightly alkaline (> 7) and as low as possible of turbidity and suspended solids.   These can interfere with the ozone oxidation process by protecting the target contaminant from ozone contact.   Post-treatment with mechanical filtration is also recommended in cases where iron and manganese have been oxidized to insoluble forms.   GAC post-treatment can also be used to remove any partially oxidized organic contaminant from the water.

Chlorine

Although it can oxidize some compounds, chlorine is mainly used for disinfection.   Therefore, it is necessary to produce and maintain adequate chlorine residuals.   Chlorine solutions will lose their potency upon standing or when exposed to air or sunlight.   Fresh chlorine solutions should be made frequently to maintain necessary residuals.   In order to control microbiological contaminants, it has been recommended that one maintain a free chlorine residual of 0.3 to 0.5 mg/L after a 10 minute contact time for household systems.[19]   This chlorine residual should be measured frequently to ensure that all systems are operating correctly. If the water source shows an increased chlorine demand and additional chlorine has been added to compensate for this demand, the chlorine dose should not be decreased without frequent measurements of chlorine residual.

Liquid chlorine solutions are slightly more stable than solutions from dry chlorine but when properly stored, the dry powder is stable.   All chlorine solutions must be protected from sun, air, and heat.   The dry chlorine powder should never be stored near flammable materials.   When dissolved in water, the chlorine powder produces a heavy sediment that can clog equipment unless filtered.

Another concern with using chlorine is the production of chlorination by-products.   Chlorine will react with naturally occurring organic matter such as humic and fulvic acids to produce chlorination by-products.   Some of these by-products such as the trihalomethanes have been considered a health risk and their concentrations have been regulated by the U.S. Environmental Protection Agency.   Currently, the maximum contaminant level for total trihalomethanes in drinking water from community systems is 0.10 mg/L.   By 1992, new disinfection by-product regulations could lead to lower allowable total trihalomethane concentrations and regulations for other disinfection by-products such as haloacids and haloacetonitriles. One way to help control the production of these chlorination by-products is to remove as much of the humic material (precursors) as possible before chlorination.   This can be done by granular activated carbon filtration or by membranes.

Iodine

Although iodine appears to be an effective biocide, its use is limited because of a concern that residual iodine in the water may be a problem for people with thyroid conditions. The only currently approved use of iodine disinfection is a traveler's pour-through cup which must carry a label stating that it is for emergency use only.[20] Therefore, there is little operational information on these units.

MAINTENANCE

A well-defined maintenance program is essential for providing continuous quality water. This maintenance program may consist of an on-demand contract with a local plumbing contractor, a point-of-use/point-of-entry service representative or dealer, a water service company, or a local water utility. Also, for a limited time period, equipment maintenance may be provided as part of an installation warranty. Maintenance should always be performed by personnel familiar with the installations.

Water quality and water use vary between homes so that the servicing of granular activated carbon units before exhaustion is critical. Monitoring records will prove invaluable toward establishing regular servicing increments. The use of water meters, although adding to the cost, will greatly assist this task.

Another method of servicing each type of granular activated carbon unit would be to replace cartridges, carbon, and so on before their respective exhaustion. Two methods are available: using a strict calendar replacement (e.g., every six months); or using flow shutoffs or alarms to signal when a certain gallonage has been treated. The latter types requires the customer to call for service, but the alarm can be preset conservatively to allow adequate time for response.

Dealers or homeowners can easily replace cartridge units, whether they are exhausted carbon media or the self-contained cartridge. The point-of-entry units contain larger volumes of carbon and will require a serviceman to replace the unit. The dealer or serviceman will either collect the carbon for batch regeneration or discard the carbon.

Typical maintenance requirements for membrane systems include periodic cleaning or replacement of the membrane. The frequency of maintenance is usually determined by raw water quality characteristics but, in general, membranes should only require replacement every 2 to 3 years. For ion exchange systems, the major maintenance requirements is periodic regeneration of the resin. With use, however, resin capacity will be reduced because resin effectiveness is never completely recovered by regeneration. When the capacity becomes too low, the resin can be washed with either a strong acid or base solution or it can be replaced. Oftentimes, replacement is more feasible.

For a distillation system, typical maintenance requirements include periodic cleaning of the evaporator and storage tanks and replacement of the heating element. Cleaning frequency which ranges from monthly to annually will depend on the hardness and total dissolved solids of the raw water. Heating elements should typically last for at least 3 years before replacement.[9]

With air stripping, blower up-keep is the most significant maintenance requirement for most systems. This normally consists of periodically lubricating the motor and changing the air filter. Other maintenance requirements may include those associated with any pumping and storage equipment included in the system. Also, if high iron or manganese is in the raw water, periodic cleaning of the packing material may be required.[9]

With activated alumina treatment, the major maintenance requirement is the periodic replacement or regeneration of the activated alumina media. With ultraviolet treatment, periodic cleaning of the disinfection chamber and lamp replacement are required. Some systems have automatic or manual mechanical cleaners for cleaning the lamp tubes without disassembling the unit. The frequency of cleaning will depend on influent water characteristics. Typical lamp replacement is annually.[9]

For ozone treatment systems, periodic cleaning of the storage tank and replacement of the air desiccant is required.[15] Other requirements may include normal maintenance of pumps, fans, and valves. These same maintenance requirements are needed for chlorination systems.

MONITORING

A monitoring program should be designed to assure that the equipment is functioning properly and to check the quality of the raw and treated water. Equipment installed in a community with water contaminated by chemicals potentially hazardous to health should be monitored more frequently than those installed in areas where contamination is of aesthetic concern. Analyses of samples for organics, inorganics, microorganisms, and radionuclides should be done by a certified laboratory.

A compliance monitoring program should be developed to evaluate and demonstrate the effectiveness of the units, provide data on the quality of the water consumed by residents of the service area, and test different types of equipment against different water quality conditions.

Compliance monitoring for a water district of all installed equipment is not necessary. For example, a percentage of granular activated carbon unit installed for each chemical should be sampled monthly. During initial testing samples should include raw and treated water. Representative households with high levels of contamination and typical installations should be selected. A data base will develop that shows the length of time a device can operate before exhaustion. The monitoring frequency can then be adjusted based on these results.

A general surveillance monitoring program should also be implemented that will satisfy several objectives: (1) to respond to residents' questions about raw water quality at homes without treatment, (2) to assess trends in the quality of raw water, and (3) to respond to questions from homeowners with point-of-use treatment devices.

REFERENCES

1. Point-of-Use Carbon Treatment Systems, Albany, NY: Division of Environmental Protection, New York State Department of Health, (1982).

2. Baier, J. H. and T. Martin, "Carbon Filter Sanitation", Suffolk County Department of Health Services, Suffolk County, NY, (1983).

3. Hand, D. W., Crittenden, J. C., Miller, J. M., and Gehin, J. L., "Performance of Air Stripping and GAC for SOC and VOC Removal From Groundwater", EPA/600/S2-88/053, (January, 1989).

4. Lykins, Jr., B. W., Goodrich, J. A., and Clark, R. M., "POU/POE Devices: Availability, Performance, and Cost", Presented at ASCE National Conference on Environmental Engineering, Austin, TX, July, 1989.

5. Lowry, J. D., "Radon Progeny Accumulation in Field GAC Units", Final Report, Maine Department of Human Services, Division of Health Engineering, (March, 1988).

6. Montemarano, J., "The Activated Carbon Disposal Dilemma", *Water Technology*, (August, 1988).

7. Moran, D., "Granular Activated Carbon Treatment Units for Removal of Aldicarb Residues in Private Wells of Suffolk County", Suffolk County Department of Health Service, Suffolk County, NY, (1983).

8. Montemarano, J. and R. Slovak, "Factors That Affect RO Performance", *Water Technology*, (August 1990).

9. "Guide to Point-of-Use Treatment Devices for Removal of Inorganic/ Organic Contaminants from Drinking Water", New Jersey Department of Environmental Protection, (1985).

10. Slovak, J. and R. Slovak, "Reverse Osmosis Checklist -- Things You Should Know Before Installing RO", *Water Conditioning and Purification*, (May, 1983).

11. "Water Treatment Plant Design For the Practicing Engineer", edited by R. L. Sanks, Ann Arbor Science Publishers, Inc., Ann Arbor, Michigan, (1978).

12. "Survey of Test Protocols for Point-of-Use Water Purifiers", Ottawa, Canada: Department of National Health and Welfare, (August, 1977).

13. McGarvey, F. X. and S. M. Ziarkowski, "Choosing the Right Resin", *Water Technology*, (September, 1990).

14. "Home Water Treatment: What's the Use of Point-of-Use?", *Health and Environment Digest*, 3(6):1-3, (July, 1989).

15. "Distillation For Home Water Treatment", Cooperative Extension Service, Michigan State University, East Lansing, MI, (January, 1990).

16. "Laboratory Testing of Point-of-Use Ultraviolet Drinking Water Purifiers", Ottawa, Canada: Health and Welfare, (April, 1990).

17. Carrigan, P., "Water Disinfection Using Ultraviolet Technology", Proceedings of the Water Quality Association Annual Convention, San Antonio, TX, March, 1990.

18. Lykins, Jr., B. W., J. A. Goodrich, and J. C. Hoff, "Concerns With Using Chlorine Dioxide Disinfection in the USA", *Journal of Water Supply Research and Technology -- Aqua*, 39(6):376-386, (December, 1990).

19. Andrews, S., "P.O.E. Ozone Systems -- Standard Methods and Materials of Application", *Water Conditioning and Purification*, (August, 1990).

20. Wagenet, L. and A. Lemley, "Chlorination of Drinking Water", Cornell Cooperative Extension, New York State College of Human Ecology, Fact Sheet 5, (September, 1988).

21. Montemarano, J., "Demand-Release Polyiodide Disinfectants", *Water Technology*, (August, 1990).

CHAPTER 5

INTRODUCTION

In attempting to solve one water quality problem by the use of POU/POE technologies, other secondary health effects can result because of disinfection by-products, contaminant breakthrough, microbiological breakthrough, radiation exposure, sodium sensitivities, and bypass of units entirely. Each POU/POE technology described previously in Chapter Two can in some way create new health problems of concern. Prevention and/or control of these problems can be accomplished through proper operation and maintenance of the devices that were discussed in detail in Chapter 4. The following sections describe the health concerns of each of the POU/POE technologies.

GAC

Microbiological Concerns

Bacterial growth in activated carbon units has drawn much attention from researchers and regulators alike. The bacteria in finished drinking water can originate by passing through the treatment plant in flocs or turbidity, being entrapped in crustacea and nematodes, come from bird droppings in open reservoirs, biological growth in tanks, or sediment or biofilm pockets in distribution systems. The organic material adsorbed on the carbon provides food for bacterial growth, and stagnation periods between operation of the units allow time for bacteria to colonize. Carbon particles may be densely colonized and problems may arise when particles enter finished drinking water.[1-8]

As the filter is used, water passes through the activated carbon bed and some of the organic material in the drinking water is concentrated on the surface of the activated carbon particles within the bed. Growth is favored at warm room temperatures, low flow rates, and periods of stagnation (e.g. overnight or longer). This is of concern because total bacterial numbers (SPC or HPC) can be elevated, coliform bacteria can grow on the filters, and pathogenic bacteria may colonize and proliferate.

173

The most common acute health effect from microbiological contamination is gastroenteritis. Other effects include headaches, stomach cramps, vomiting, diarrhea, fatigue, and nausea. Viruses can cause meningitis and paralysis. Coliforms, although not necessarily disease producing, can be indicators of other organisms that can cause dysentery, hepatitis, typhoid fever, and cholera.[9] Secondary use (washing), most likely to be associated with POE filters, may promote pulmonary infections because of the generation of aerosols either by showerheads or humidifiers. The most prominent respiratory disease mentioned in the literature is *Legionella*, but other bacteria such as *Acinetobacter* and *Pseudomonas* have been found in potable water isolated from GAC filtered effluent.[10-13] Of the three million cases of pneumonia in the United States each year, only in one-third of the cases is the causative agent identified. Thus, raising the question of the role potable water systems and GAC filters may play in some of these cases.[14]

The bacterial quality of water produced by home treatment devices will differ from that delivered to the faucet by the public utility.[15-17] Wallis et al.,[15] were among the first to demonstrate excessive growth of bacteria on activated carbon filters. They found bacterial levels of up to 70,000/mL within 6 days of installing a new filter. Fiore and Babineau[18] obtained high levels of bacteria in the first flush after overnight stagnation, but maintained that these were not significantly higher than influent levels after periods of stagnation. Two other studies during the 1970's[19,20] showed high total bacterial counts after the filters had been in use for several days. Studies carried out by the Canadian Bureau of Chemical Hazards on several types of activated carbon devices have shown elevated counts in the effluents from the devices on both raw and finished drinking water.[21] Tobin et al.,[22] reported on simulated lifetime testing of three activated carbon devices. During the first few days in service, the filters were effective in removing bacteria from water, but effluent counts gradually increased until they reached a plateau at about 25 days. With a chlorinated drinking water containing a background level of 160 colonies/mL, the first samples taken from the three devices after overnight stagnation contained geometric means (days 25-55) of 85,000, 40,000 and 162,000 colonies/mL. Table 1 shows typical counts from a carbon unit and the importance of flushing the tap. Bacterial densities in the product water from POU GAC filters can be expected to increase by 1 to 2 logs over the number detected in public water sampled at the tap.[16] Microbial proliferation relates to the species present in the tap water influent to the filter device, the presence of free-chlorine residual, seasonal changes in water temperatures, ambient air temperatures around the device, and the service duration for the given unit.[23] Organisms present in the well or passing through the distribution system may become transient colonizers in a carbon filter for varying periods of time.[24] Table 2 lists organisms isolated from GAC in several studies.

TABLE 1.   STANDARD PLATE COUNTS
IN GAC EFFLUENT STREAM[25]

TIME	SPC/mL
First Flush	7,800
1 Minute	50
3 Minutes	20

TABLE 2.   BACTERIA ISOLATED FROM GRANULAR ACTIVATED CARBON

Bacterial
Genus

Citrobacter	Alcaligenes
Klebsiella	Achromobacter
Serratia	Xanthomonas
Proteus	Moraxella
Aeromonas	Bacillus Species
Pseudomonas	Corynebacteria
Acinetobacter	Staphylococcus
Flavobacterium	Enterobacter

This list should not be interpreted as complete or exhaustive.   It
is likely that many bacterial strains or species capable of adsorbing to
GAC may go undetected because of inadequate isolation media or procedures.
The list is instructive in that it indicates that a wide variety of bacteria adsorb to GAC.[14]

A majority of the bacteria found in drinking water and on GAC are
native to aquatic environments and not normally considered to be pathogens.   However, endotoxins have been found in measurable quantities and
should be studied further.

The Canadian studies showed standard plate count (SPC) values elevated after filtration, and two pathogens, *P. aeruginose* and *Flavobacterium*
were identified.   *Pseudomas aeruginosa* was present in the effluent at a
level of 3-6 colonies/100 mL and were not detectable in the influent municipal treated water.   In a study of three brands of filters[22], *P.
aeruginose* were isolated from the effluents of each.   It is noteworthy
that growth of coliforms and *Pseudomas* on hydrocarbon-exhausted activated

carbon has been used as a method of bioregeneration.[26]   Several outbreaks of gastrointestinal illnesses have been attributed to drinking water contaminated with *P. aeruginosa*.[27]   It is also the cause of persistent ear and urinary infections[28,29].   Seyfried and Fraser[30] determined that *P. aeruginosa* persisted in swimming pool water containing 0.4 mg/L free available chlorine and also demonstrated ear infection in 4 of 24 people swimming in a pool containing less than 80 organisms/100 mL[31].   In hospitals, this organism is particularly dangerous for those who are immunologically impaired, burn cases, and those undergoing surgery.   It has caused epidemics of diarrhea in a nursery for newborns in which several children died.[32]   *Flavobacterium* is also a pathogen of some importance for newborns.   It has been identified as the pathogenic agent in an epidemic of meningitis of newborns[33] and is a pathogen for patients undergoing surgery as well.[34]   In one hospital study[35], the upper airway of 195 patients were colonized by *Flavobacterium* that were ultimately traced back to the water supply.   In considering pathogens, it is well established that there is a minimum infective dose for each pathogen below which very few persons will become ill.   In most cases, sporadic or transient contamination of drinking water will result in concentrations below this dose.   It is the possibility of exacerbation, by growth of this organism on filters, that is a cause for concern.   Some persons such as the aged and very young infants may also have significantly lower resistance to the pathogens.   The many uses of water such as preparing baby foods and formula, humidifying sick-rooms, washing wounds, bathing and showering, all present different opportunities for infection to occur.

The extent of colonization by pathogens is limited by the native microbial community.   Camper et al.,[36] found that pathogens could survive for extended periods of time when virgin carbon was used.   A well developed biofilm of indigenous organisms compete for nutrients, space, and produce secondary metabolites that limit pathogen growth.   Thus, pretreatment (chlorination, ozonation, etc.) is very important following unit installation or the filter could be left out of operation for one to two weeks to allow the establishment of a microbial community.   This enhanced growth may extend the period of time before the GAC must be regenerated.[37]   On the other hand, if opportunistic pathogens are present in this microbial flora, their growth may reach an infective dose level for some individuals when amplification of the heterotrophic bacterial population occurs (over 1,000 organisms per ml).   Muller[38] has documented cases when disease outbreaks have followed sudden increases in plate counts. High bacterial counts may also cause the loss of coliform test sensitivity, thus providing a false sense of security.   Potential for adverse health affects to the general population from ingestion of large numbers of heterotrophic bacteria in the product water appears to be low.   Obviously, epidemiological studies are needed on water supplies with high heterotrophic bacterial populations. Even well-designed studies may have difficulty identifying cause and effect because of the apparent randomness in sloughing of carbon particles and microbial populations.   Routine local health department records provide little information because of the variation in symptoms and severity of gastrointestinal illness that would not be reported.

Other studies have shown that bacteria attached to carbon particles passing through GAC filters can also be an important source of contamination in drinking water.[39] Results showed that the accumulation of carbon fines was dependent only on the amount of water filtered and not on the time period before or after filter backwash. This indicates that release of carbon fines could be a random event not related to filter operation. Bacteria attached to GAC have also shown to be unaffected by disinfection with 2 mg/L of free residual chlorine for one hour.[40] Operational variables indicate that increased bed depth, higher applied water turbidity, and increased filtration rate resulted in either higher numbers of released particles, an increased heterotrophic plate count (HPC) bacterial load on the particles, or an elevated number of attached coliforms.[25]

Silver, or some other bactericide, has been incorporated in some units in an attempt to control bacterial activity.[22,41] Since 1975 there have been over 140 registrations issued by EPA for bacteriostatic media and filters which incorporate a pesticide type chemical such as silver to prevent build-up of bacteria in the filter bed. EPA registration is now given to a product after acceptable evidence is submitted to validate that the labeling claims are true, and that it does not impart over 50 μg/L of silver to the treated water.[42]

Several documents[21,42] have reported the inactivation of total coliform organisms, including *Escherichia coli*, by silver impregnated units. However, the general bacterial population, possibly including pathogenic bacteria, was not affected. The effectiveness of silver in controlling the bacterial population is apparently limited, because bacterial densities have been shown to be as high as those on units that did not contain silver. A paired silver-containing unit actually had higher bacterial numbers in the effluent than the non-silver containing control. The United States Environmental Protection Agency commissioned a study on activated carbon units and showed similar results. The seven brands of filters yielded maximum bacterial values on first flush samples of 190,000, 120,000, 320,000, 240,000, 20,000, 120,000 and 130,000 per mL.[43] Among these, devices containing silver did not consistently effect bacteriostasis in their tests. The bacterial flora of the carbon impregnated with silver appeared to be different in colony appearance and types of organisms recovered compared to that of carbon filters without silver. Apparently, silver serves as a selective agent, allowing silver tolerant bacterial strains to grow.[19] If silver is to be used, it should be monitored to ensure that the product water does not exceed acceptable limits, thus creating another health problem. Most carbon block units do not contain silver, but claim bactericidal effects, probably due to straining through small carbon pore size (0.5u).[44]

Chemical Concerns

Substantial competition for adsorption sites by the contaminants of concern, other organics in the source water, chlorine, and the microbial community can lead to earlier than expected breakthrough of detectable amounts of contaminants from the carbon bed. Large hydraulic loadings, moderate to poor adsorption on the GAC by the contaminant or contaminants of interest, and significant exhaustion of the GAC bed can also lead to early breakthrough as described in Chapter 4. Chemical contamination from POU/POE devices themselves have been documented for methylene chloride [45], ethylene oxide [46], and dimethylnitrosamine.[47]

Radiation Concerns

Radon is known to cause lung cancer when inhaled over a long time. Waterborne radon is a problem only when the radon is released from the water and enters household air. Drinking water containing radon is not considered a health risk. Scientist estimate that approximately 100 to 1,800 lung cancer deaths per year are caused by inhaling radon emitted by household water. The lifetime risk of developing lung caner from household water that contains 1,000 pCi/L of radon is roughly 3 to 13 in 10,000; from water with 10,000 pCi/L of radon, the risk is approximately 3 to 13 per 1,000; and for water containing 100,000 pCi/L of radon, the risk is about 3 to 12 per 100.[48]

GAC is essentially unlimited in its ability to remove radon because radon decays into other radioactive products while being held in the GAC tank. However, elevated gamma exposure levels and build-up of lead are possible in GAC beds when being used to remove radon-222. The progeny of 222 Rn make it important to consider shielding for the GAC bed or limited access to the unit (see Chapter 2).[49]

MEMBRANE FILTRATION

Microbiological Concerns

Various investigators have noted that membranes have imperfections that will preclude complete rejection of bacteria and viruses.[50,51,52] Inorganic fouling, bacterial attack, temperature, pH, and disinfectants can affect membrane performance and life, thus potentially compromising health in the same manner described previously on GAC microbiological concerns. Enrichment of bacteria in membrane unit components (holding tank, membrane surface) can result from periods of non-use as brief as overnight or weekends. The worst product water is usually found during the first flush as is found with GAC units.

The environment offered by RO membranes (high surface area, always moist, room temperatures) provides excellent conditions for bacterial growth. Attached biofilms can trap dissolved and suspended materials to serve as bacterial nutrients. These biofilms in addition to fouling the membrane may shield bacteria from disinfectants. Permeate quality may deteriorate as a result of "concentration polarization". This is increased salt passage which is due to the entrapment of dissolved solids in the biofilm layer.

Although virtually all reverse osmosis membranes have pore sizes too small to allow bacteria to penetrate the membrane layer, bacteria can appear on the permeate side of the membrane. This is a relatively little known phenomenon called "grow-thru". The rapid movement of water over the membrane surface of an operating pump-driven unit will minimize bacterial grow-thru. But line pressure point-of-use devices and pump driven units that are sitting idle will be affected.

Another source of bacterial contamination is the RO system pretreatment device. Prefilter cartridges provide a high surface area media covered with nutrients. Although an activated carbon filter is used to remove the most common chemical disinfectant, chlorine, contact occurs only in the upper layers, while the rest of the cartridge affords an excellent growing environment for bacteria.

It is important to understand that many of the pretreatment components necessary to maximize the life of the reverse osmosis membrane may actually contribute to the bacterial fouling problem if an adequate disinfection regiment is not implemented.[50]

During the course of an epidemiological study on health hazards associated with drinking water, 300 RO water filtration units were analyzed for bacterial content.[51] The RO POU devices produced 2 L/h of water that was stored in a 10 L pressurized bladder reservoir. The RO system consisted of a sediment prefilter, an activated charcoal filter, and thinfilm/polymer Amarid membrane.

Analysis by the pour plate method of 300 samples revealed that water from the RO units contained 10^4 to 10^7 cfu/mL (Figure 1). When 130 of these units were retested approximately 4-6 months later, most units delivered water containing 10^3 to 10^5 cfu/mL (Figure 2), values generally lower than those observed during the first sampling period. The results at 20°C and 35°C were similar.

The weekly samples obtained from some of these units revealed that the bacterial contamination was fairly stable over time (Figure 3). Attempts to reduce the bacterial counts in the reservoirs using chlorine bleach were not successful: they returned to their high value within a few days. Chlorine could not be used on the RO membrane itself because these membranes are highly susceptible to chlorine. The membrane efficiency was not affected by this bacterial growth: all units produced water meeting the conductivity standard set for such units.

The units appeared to be colonized by only one species at a time. The appearance on R2A medium of the bacteria was very constant and colonies were beige, yellow, or more rarely, pink pigmented. The bacteria isolated were mostly Gram-negative, rods or cocci, oxidase positive or negative. Speciation revealed the presence of *Pseudomonas*, *Acinetobacter*, *Flavobacterium*, *Chromobacterium*, *Alcaligenes*, and *Moraxella* (Table 3).

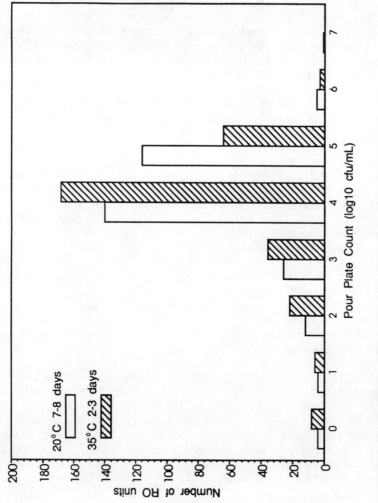

Figure 1:   Correlation plot and histogram of results obtained from RO units reservoir water samples by the pour plate method, as determined by incubation at 20 or 35^0 C on R2A medium.

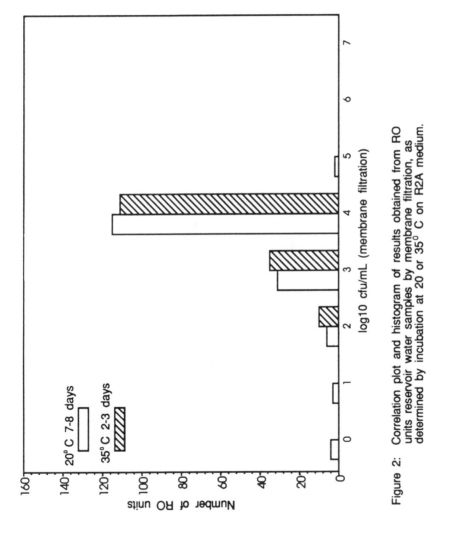

Figure 2:   Correlation plot and histogram of results obtained from RO units reservoir water samples by membrane filtration, as determined by incubation at 20 or 35° C on R2A medium.

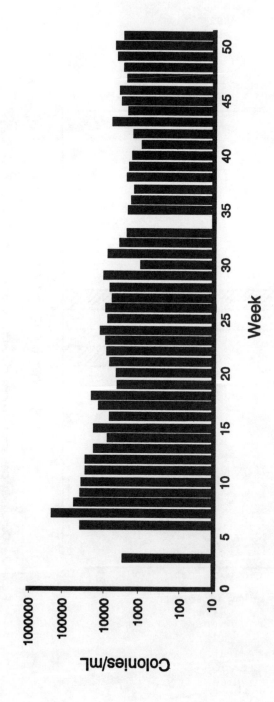

Figure 3: Number of colonies per millilitre, obtained over a 1-year period from an RO unit, as determined by incubation at 20° C on R2A medium.

Bacteria isolated at 20°C do not appear to belong to different species than those isolated at 35°C. The bacterial species found in these water samples are similar to those reported by other researchers for activated charcoal filters and domestic water filtration units.[17,41]

TABLE 3.   IDENTIFICATION OF 25 GRAM-NEGATIVE BACTERIA
ISOLATED ON R2A MEDIUM FROM RO UNITS[51]

Unit	Bacteria identified
1	*Acinetobacter calcoaceticus*
2	*Alcaligenes* or *Moraxella* spp.
3	*Alcaligenes* or *Moraxella* spp.
4	*Chromobacterium* spp.
5	*Flavobacterium* spp.
6	*Flavobacterium* spp.
7	*Pseudomonas* spp.
8	*Pseudomonas* spp.
9	*Pseudomonas* spp.
10	*Pseudomonas* spp.
11	*Pseudomonas* or *Alcaligenes* spp.
12	*Pseudomonas* or *Alcaligenes* spp.
13	*Pseudomonas* or *Alcaligenes* spp.
14	*Pseudomonas* or *Alcaligenes* spp.
15	*Pseudomonas* or *Alcaligenes* spp.
16	*P. fluorescens* or *P. putida*
17	*P. paucimobilis*
18	*P. paucimobilis*
19	*P. paucimobilis*
20	*P. paucimobilis*
21	*P. paucimobilis* or *maltophilia*
22	*P. paucimobilis* or *maltophilia*
23	*P. paucimobilis* or *maltophilia*
24	*P. paucimobilis* or *maltophilia*
25	*P. paucimobilis* or *maltophilia*

Results suggest that the heterotrophic bacterial population in water collected from the reservoirs of these domestic RO units is very homogeneous. When one bacterial specie has colonized the reservoir, or most probably the RO membrane, only this specie is present. Very large numbers of bacteria have been detected in the water from these units and the average counts are several orders of magnitude above those of tap water. The bacterial species detected themselves are not recognized pathogens through ingestion and should thus not pose any particular health risks when found in drinking water, however as in GAC units, elevated plate counts can protect and aid opportunistic pathogens allowing them to reach infective levels. There may be aesthetic incentives as well to keep these concentrations in check.

Chemical Concerns

When utilizing membrane filtration to remove health related contaminants such as pesticides or nitrates, it is very important to utilize the correct membrane. Nitrates for example, are poorly rejected by cellulosic membranes. Only thin film composite (TFC) should be considered. The specific health-related ion should be regularly monitored in addition to using built-in TDS monitoring or the preferred percent rejection monitor.[52]

Both nitrates and pesticides are contaminants which are a concern in groundwater. Nitrates are reduced to nitrites in the body. Infant methemoglobinemia is probably the best documented health concern from nitrites. Very high concentrations of nitrates in drinking water can be fatal to infants, particularly within the first three months of life.[53] Pesticide ingestion has been associated with health problems that include cancer, nervous system disorders, birth defects, and male sterility.[54]

ACTIVATED ALUMINA

Microbiological Concerns

Coliform bacteria levels do not typically increase throughout the filter bed and have even been observed to decrease. Noncoliform bacteria growth on the media has been observed and can result in the same problems associated with high plate counts found on other filter media. Once again, flushing the tap significantly reduces the levels.

AIR STRIPPING

Microbiological Concerns

Packed tower aeration units are often equipped with post-treatment disinfection, usually chlorination or UV light. There is little evidence documented regarding the development of biofilm on the packing material in drinking water applications. However, some packed towers designed for hazardous waste treatment depend on some level of biological degradation of the toxic waste through biofilm on the packing material. Several large-scale community towers install piping to accommodate chlorine injection into the top of the tower should it become necessary to shock the packing and remove the biofilm either because of microbiological concerns or excessive pressure drop. POE units are often designed to incorporate UV treatment following GAC filtration prior to distribution to the household.

Chemical Concerns

As mentioned previously in Chapter 4, aeration can produce a more aggressive water that could cause or increase the rate of corrosion of water heaters, plumbing, and joint materials, thus subjecting consumers to excess levels of lead, copper, iron, etc. and the associated health effects described in Chapter 1.

DISTILLATION

Microbiological Concerns

Most bacteria in the raw water are killed by the high temperatures and settle to the bottom of the boiling tank. Those that are not killed are separated from water as the steam rises from the tank. One possible risk of microbiological contamination could arise from airborne contaminants being deposited in the finished water when allowed to be in contact with the open air in order to improve the taste by dissolving oxygen and carbon dioxide.

Mineral Removal

There is the fear that the "good" minerals in drinking water are removed through distillation and therefore human health will suffer. Drinking water accounts for about 5 percent of the ingested minerals with food accounting for the remaining 95 percent.[56] This is a another consideration the consumer will have to balance when contemplating purchase of a POU/POE device.

Chemical Concerns

Carryover of volatile organic contaminants are a primary concern of distillation treatment. Contaminants that vaporize near to the boiling point of water can be collected on the condenser and actually be concentrated in the treated water. Units with volatile gas venting systems, pre-heat chambers, or post-treatment GAC cartridges have been produced to counter this problem.[56]

ION EXCHANGE

Microbiological Concerns

Bacterial growth in ion exchange units is common and effluent levels can exceed raw water numbers. The phenomenon is similar to other filter media (GAC, activated alumina). The concentrations tend to be highly variable but not significantly different from the influent. Once again, concerns of high HPC levels providing protection for opportunistic pathogens arises. As with the other treatment systems mentioned, regeneration, backwashing, and/or flushing can lower bacterial numbers in the filter beds. However, post-disinfection should still be required to ensure consumer protection.

Chemical Concerns

In cation exchange the product water can have elevated sodium levels. Chronic excessive sodium levels has been associated with hypertension and increased blood pressure.[57]   Since these home water softeners may pose some health risks for people with high blood pressure and heart disease, they should check with their physicians to determine whether this treated water will aggravate their health problems. The amount of sodium in water softener effluent depends on the hardness of the water. For example, one grain of hardness/gallon (64.7 mg/gallon) of raw water adds 30 mg sodium/ gallon of softened water.   Recent studies question the relationship between sodium and hypertension as well as the contribution of drinking water as a source of sodium relative to dietary intake.[58]   Nevertheless, post-treatment with RO is an option to reduce sodium to lower levels for those persons with high blood pressure and on a medically supervised sodium restricted diet.

Product water from anion exchange units may become corrosive to the household plumbing and result in elevated levels of copper, iron, or lead in addition to shortening the life of the plumbing and fixtures. Selective contaminant removal is another concern with anion exchange. Sulfate is preferred by most anion exchange resins over nitrates. This will reduce the nitrate exchange capacity of the resin as well as possibly release previously exchanged nitrate ions back into the effluent exposing the consumer to a slug of nitrates.[59]

DISINFECTION AND OXIDATION

Ozone Microbiological Concerns

Since ozone decomposes organic matter in such a way that it becomes more biodegradable and ozone residuals are short lived, bacterial regrowth can be a problem in distribution systems. This problem has of yet not been documented in household plumbing. It is possible that the opposite problem may occur in POU/POE applications in that ozone residuals may remain in the water because of too short a detention time between treatment and consumption of the water.

Ozone Chemical Concerns

Breathing traces (< 0.1 ppm) of ozone in the air for a few minutes is not a major health problem. However, breathing higher concentrations for prolonged periods of time can produce various physiological responses.[60]

The odor threshold of ozone varies among individuals, but most people can detect 0.01 ppm in air. This is well below the level for general comfort. Symptoms experienced when humans are exposed to ozone levels of 0.1 to 1 ppm are headache, dryness of throat, irritation of the eyes, and burning of the eyes. Exposure to 1 to 100 ppm of ozone can cause asthma-like symptoms, e.g., tiredness, nausea, lack of appetite. Short term exposure to higher concentrations can cause throat irritations, hemorrhaging, and pulmonary edema.

The symptoms of ozone exposure are acute, that is, there appear to be no chronic effects among normal, healthy people. This is because the human body has the ability to repair such damages. However, this is not the case with prolonged exposure by persons having chronic health problems, or when experiencing prolonged exposure to higher concentrations.

During more than 80 years of commercial use of ozone, there has never been a case reported of exposure to ozone which has resulted in death.[60]

Ozone Disinfection By-products

Originally, ozone was thought to be a superior alternative to chlorination because of its lack of THM formation. However, ozonation by-products have been identified and could present the same concerns as THMs that have already been regulated. Partial oxidation of organics mentioned above can also produce daughter compounds that may have worse health effects than the parent compound. Table 4 list the compounds that will likely be present in finished water after "moderate-to-exhaustive" ozonation.[61]

TABLE 4.   COMMON OZONATION BY-PRODUCTS[61]

Compound	Structure	CAS registry #
oxalic acid	$HOOC-COOH$	144-62-7
mesooxalic acid	$HOOC-C(OH)_2-COOH$	473-90-5
glyoxylic acid	$HOC-COOH$	298-12-4
glyoxal	$HOC-COH$	107-22-2
methyl glyoxal	$HOC-COCH_2$	78-98-8
formic acid	$HCOOH$	64-18-6
formaldehyde	$HCHO$	50-00-0

Table 5 illustrates some specific precursor organic chemicals which, when ozonated, will produce the compounds listed above in Table 4.

TABLE 5.   OZONATION OF SOME SUBSTITUTED AROMATIC AND
ALIPHATIC CARBONYL/CARBOXYLIC COMPOUNDS.   INITIAL
CONCENTRATION 1 MMOLE/L; OZONE DOSE
10 MG O_3/MIN/L; OZONATION TIME 20-180 MIN[61]

COMPOUNDS	OXIDATION PRODUCTS
trans, trans-muconic	formaldehyde, glyoxal, glyoxylic acid, oxalic acid, formic acid, CO_2, H_2O_2
formic acid	CO_2, H_2O
glyoxylic acid	oxalic acid, CO_2
maleic acid	formic acid, glyoxylic acid
fumaric acid	formic acid, glyoxylic acid, oxalic acid, mesoxylic acid, aldehyde, CO_2
glyoxal	glyoxylic acid, oxalic acid, CO_2
tartronic acid	mesoxalic acid
malonic acid	tartronic acid, mesoxalic acid, CO_2, H_2O_2
dihydroxyfumaric acid	oxalic acid, hydroxytartaric acid, CO_2
oxaloacetic acid	formic acid, glyoxylic acid, oxalic acid, mesoxalic acid, CO_2
p-toluenesulphonic acid	methylglyoxal, acetic acid, pyruvic acid, formic acid, oxalic acid, CO_2, H_2O_2, H_2SO_4
2-nitro-p-cresol	methylglyoxal, glyoxylic acid, acetic acid, pyruvic acid, formic acid, oxalic acid, CO_2, H_2O_2, HNO_3
4-choro-o-cresol	methylglyoxal, acetic acid, pyruvic acid, formic acid, oxalic acid, CO_2, HCL, H_2O_2

Ultraviolet Irradiation (UV) Microbiological Concerns

UV treatment systems do not maintain any residual disinfection capabilities and this may or may not be a concern.   In most POU/POE applications this should not be an issue.   The major health concern with UV utilization is ensuring the treatment unit is operating optimally as described in Chapter 4.

Chlorination Microbiological Concerns

Chlorination offers the advantage of continued disinfection after initial treatment that ozone and UV do not.   However, as previously reported, there are chlorine resistant organisms and organisms protected by particulates or biofilm that can enter finished water when sloughed off filter beds or membranes.

Chlorination Chemical Concerns

Most home applications of chlorination are at the well.   When chlorine gas is liberated from a chlorine cylinder, or moistened crystals or pellets, the fumes can be dangerous and even lethal.   Trihalomethanes have probably been one of the most highly researched group of contaminants in history.   Table 9 (Chapter 1) lists the disinfection by-products that may be regulated because of potential long-term carcinogenic health effects.

USEPA is currently collecting information on the carcinogenic health effects of the by-products listed earlier.   Information available on the carcinogenicity of some of the DBPs is summarized in Table 6.

An MCLG of zero will be established for those DBPs determined to be probable human carcinogens.   A MCLG will be established for those determined to be possible human carcinogens based on noncarcinogenic effects or in the one in ten thousand to one in a million risk range.

TABLE 6.   CARCINOGENIC HEALTH EFFECTS
OF DISINFECTION BY-PRODUCTS

Probable Human Carcinogens	10-4 Risk Level	10-6 Risk Level
Chloroform (v)	600 μg/L	6 μg/L
Bromidichloromethane (T)	30 μg/L	3 μg/L
Dichloroacetic Acid (T)	NA	NA
Trichloroacetic Acid (T)	NA	NA
2,4,6-Trichlorophenol (V)	300 μg/L	3 μg/L
Formaldehyde (T)	NA	NA
Possible Human Carcinogens		
Dibromochloromethane (T)	UR	UR
Bromoform (T)	UR	UR
Dichloroacetonitrile (T)	NA	NA
Dibromoacetonitrile (T)	NA	NA
Bromochloroacetonitrile (T)	NA	NA
Trichloroacetonitrile (T)	NA	NA

(T) = Tentative
(V) = Verified
NA = Not Available
UR = Under Review

REFERENCES

1.   American Water Works Association Research and Technical Practice Committee on Organic Contaminants, "An Assessment of Microbial Activity on GAC", *Journal American Water Works Association*, 73:223-224, (1981).

2.   Brewer, W. S., and Carmichael, W. W., "Microbial Characterization of Granular Activated Carbon Filter Systems," *Journal of American Water Works Association*, 71:738-740, (1979).

3.   Cairo, P. R., McElhaney, J., and Suffet, I. H., "Pilot Plant Testing of Activated Carbon Adsorption Systems," *Journal of American Water Works Association*, 71:660-673, (1979).

4.   "Activated Carbon... A Fascinating Material", Capelle, A., and de Voogs, F., Eds. (Amersfoort, The Netherlands: Norit N.V., 1983).

5.   Mattson, J. S., and Mark, Jr., H. B., "Activated Carbon", (New York, N.Y.: Marcel Dekker, Inc., 1971).

6. Parsons, F., Wood, P. R., and Demarco, J., "Bacteria Associated with Granular Activated Carbon Columns," Proceedings of the American Water Works Association Conference. Denver, Colorado. 271-296 (1980).

7. Schalekamp, M., "The Use of GAC Filtration to Ensure Quality in Drinking Water From Surface Sources," *Journal American Water Works Association*, 71:638-647, (1979).

8. Weber, W. J., Jr., Pirbazari, M., and Melson, G.L., "Biological Growth on Activated Carbon: An Investigation by Scanning Microscopy, " *Environmental Science Technology*, 12:817-819, (1978).

9. "Drinking Water Compliance Problems Undermine EPA Program as New Challenges Emerge", U.S. General Accounting Office, RCEO-90-127.

10. Shands, K. N., Ho, J. L., Meyer, R. D., et al. "Potable Waters as a Source of Legionnaires' Disease," *Journal American Medical Association*, 253:1412-1416, (1985).

11. Larsen, E. L. "A Decade of Nosocomial *Acinetobactor*," *American Journal Infection Control*, 12:14-18, (1984).

12. Cordes, L. G., Brink, E. W., Checko, P. J., et al. "A Cluster of *Acinetobacter* Pneumonia in Foundry Workers," *American International Medical*, 95:688-693, (1981).

13. Karnad, A. S., Alvarez, S., and Berk, S. L., "Pneumonia Caused by Gram-Negative Bacilli," *American Journal of Medicine*, 79(IA):61-67, (1985).

14. Dufour, A. P., "Epidemiological Concerns of Bacterial Growth in Activated Carbon Filters", Point-of-Use Water Quality Improvement Industry Technical Papers. Water Quality Association Annual Convention, Phoenix, Arizona, 22-24 (March 1985).

15. Wallis, C., Stagg, C. H., and Melnick, J. W., "The Hazards of Incorporating Charcoal Filters into Domestic Water Systems," *Water Research*, 8:111-113, (1974).

16. Taylor, R. H., Allen, M. J., and Geldreich, E. E., "Testing of Home Use Carbon Filters," *Journal American Water Works Association*. 71:577-579, (1979).

17. Geldreich, E. E., Taylor, R. H., Blannon, J. C., and Reasoner, D. J., "Bacterial Colonization Point-of-Use Water Treatment Devices," *Journal American Water Works Association*, 77:72-80, (1985).

18. Fiore, J. H. and Babineau, R. A., "Effect of an Activated Carbon Filter on the Microbial Quality of Water", *Journal Applied and Environmental Microbiology*, 34:541-546, (1977).

19.   Burkhead, C. E., McGhee, M. F., Tucker, G. J., Curtis, G. M., and Sandberg, N., "Biological, Chemical, and Physical Evaluation of Home Activated Carbon Filters", In Proceedings AWWA 1978 Annual Conference, Atlanta City, New Jersey, June 25-30, 1978.

20.   Hanes, N. B., "Lead Removal and Bacterial Growth in Home Water Purifiers", in Proceedings AWWA 1978 Annual Conference, Atlanta City, New Jersey, June 25-30, 1978.

21.   "The Hazards of Using Point-of-Use Water Treatment Devices Employing Activated Carbon", Department of Health and Welfare, Bureau of Chemical Hazards, Ottawa, Canada (1980).

22.   Tobin, R. S., Smith, D. K. and Lindsay, J. A., "Effects of Activated Carbon and Bacteriostatic Filters on Microbiological Quality of Drinking Water," *Journal Applied and Environmental Microbiology*, 41(3):646-651, (1981).

23.   Geldreich, E. E., and Reasoner, D. J., "Home Treatment Devices and Water Quality," *Drinking Water Microbiology*, G. McFeters, ed. (New York, N.Y.: Springer-Verlag, 1990).

24.   Longley, K. E., Hanna, G. P., and Gump, B. H.,"Removal of DBCP From Groundwater, Volume 1 POE/POU Treatment Devices: Institutional and Jurisdictional Factors," Water Engineering Research Laboratory, U.S. EPA, (1989).

25.   Jorgensen, J. H., Lee, J. C., and Pahren, H. R., "Rapid Detection of Bacteria", *Journal Applied and Environmental Microbiology*, 32:347, (1976).

26.   Houston, C. W., "Bio-Regeneration of Hydrocarbon-Exhausted Activated Carbon", University of Rhode Island, Kingston, RI, NTIS Document PB-288 651 (1978).

27.   Hoadley, A. W., "Potential Health Hazards Associated with *Pseudomonas Aeruginosa* in Water", A. W. Hoadley and B. J. Dutka, eds., Bacterial Indicators/Health Hazards associated with Water, ASTM STP 635, American Society for Testing and Materials, pp 80-114, (1977).

28.   Hoadley, A. W., "On the Significance of *Pseudomonas Aeruginosa* in Surface Waters", *New England Water Works Association*, 82:99-111, (1972).

29.   Whitby, J. L. and Rampling, A., *Pseudomonas Aeruginosa* Contamination in Domestic and Hospital Environments, *Lancet*, 1:15-17, (1972).

30.   Seyfried, P. L. and Fraser, D. J., "Persistence of *Pseudomonas Aeruginosa* in Chlorinated Swimming Pools", *Canada Journal of Microbiology*, 26: 350-355, (1972).

31. Seyfried, P. L. and Fraser, D. J., *"Pseudomonas Aeruginosa* in Swimming Pools Related to the Incidence of Otitis Externa Infection", *Health Laboratory Science*, 15:50-57.

32. Hunter, C. A. and Ensign, P. R., "An Epidemic of Diarrhea in a Newborn Nursery Caused by *Pseudomonas Aeruginosa"*, *American Journal Public Health*, 37:1166-1169, (1947).

33. Cabrera, H. A. and Davis, G. H., "Epidemic Meningitis of the Newborn by *Flavobacterium"*, *American Medical Association Journal Dis. Child*, 101:228, (1961).

34. Herman, L. S. and Himelsback, C. K., "Detection and Control of Hospital Sources of *Flavobacterium"*, *Hospitals*, 39:72, (1965).

35. du Moulin, G. C., "Airway Colonization by *Flavobacterium* in an Intensive Care Unit", *Journal Clinical Microbiology*, 10:155-160, (1979).

36. Campor, A. K., LeChevallier, M. W., Broadway, S. C., and McFeters, G. A., "Growth and Persistence of Pathogens on Granular Activated Carbon Filters," *Journal Applied and Environmental Microbiology*, 60:1378-1382, (1985).

37. Wilcox, D. R., Chang, E., Dickinson, K. L., and Johansson, J. R., "Microbial Growth Associated with Granular Activated Carbon in a Pilot Water Treatment Facility", *Journal Applied and Environmental Microbiology*, 46(2):406-416, (1983).

38. Muller, G., "Bacterial Indicators and Standards for Water Quality in the Federal Republic of Germany", H. W. Hoadley and B. J. Dutka, eds., Bacterial Indicators/Health Hazards Associated with Water, ATM STP 635, American Society for Testing and Materials, pp. 159-167, (1977).

39. Camper, A. K., LeChevallier, M. W., Broadway, S. C., and McFeters, G. A., "Bacteria Associated wit Granular Activated Carbon Particles in Drinking Water," *Journal Applied and Environmental Microbiology*, 52(3):434-438, (1986).

40. LeChavallier, M. W., Hassenauer, T. S., Camper, A. K., and McFeters, G. A., "Disinfection of Bacteria Attached to Granular Activated Carbon," *Journal Applied and Environmental Microbiology*, 48:918-923, (1984).

41. Reasoner, D. J., Blannon, J. C. and Geldreich, E. E., "Microbiological Characteristics of Third-Faucet Point-of-Use Devices," *Journal American Water Works Association*, 79:60-66, (1987).

42. Collins, J. D. and Polens, W. J. , "Silver ... the Ageless Bacteriocide", *Water Technology*, 11(9):32-36, (September 1989).

43.   Bell, F. A., Perry, D. L., Smith, J. K., and Lynch, S. C., "Studies on Home Water Treatment Systems", *Journal American Water Works Association*, 2:125-130, (1984).

44.   "Guide to Point-of-Use Treatment Devices For Removal of Inorganic/Organic Contaminants From Drinking Water", New Jersey Department of Environmental Protection, (1985).

45.   *Mainstream*, 33:1:6, American Water Works Association, Denver, Colorado, (1989).

46.   *Mainstream*, 31:6:1, American Water Works Association, Denver, Colorado, (1987).

47.   Simenhoff, M. L., et al., "Generation of Dimethyl Nitros Amine in Water Purification Systems", *Journal American Medical Association*, 250:14:2020-2024, (October, 1983).

48.   United States Environmental Protection Agency, "Removal of Radon from Household Water), EPA-87-011, Washington, DC 20460, (September, 1987).

49.   Lowry, J. D. and Lowry, S. B., "Modeling Point-of-Entry Radon Removal by GAC", *Journal American Water Works Association*, 79:85-88, (1987).

50.   Cartwright, P. S., "How to Choose the Right Disinfectant for Your RO System", Roseville, MA.

51.   Payment, Pierre, "Bacterial Colonization of Domestic Reverse Osmosis Water Filtration Units", *Canadian Journal of Microbiology*, 35(11):1065-1067, (1989).

52.   Montemarano, J. and Slovak, R. "Factors That Affect RO Performance", *Water Technology*, 13:12:44-54.

53.   Federal Register, National Primary Drinking Water Regulations; Synthetic Organic Chemicals, Inorganic Chemicals, and Microorganisms: Proposed Rule, Vol. 50, November 13, 1985, pp. 46935-47022, (1985).

54.   Bouwer, H., 1989, Agriculture and Groundwater Quality, *Civil Engineering*, 59(7):60-63.

55.   "Guide to Point-of-Use Treatment Devices for Removal of Inorganic/Organic Contaminants from Drinking Water", New Jersey Department of Environmental Protection, Division of Water Resources, (December, 1985).

56.   Carpenter, B., "Making the Case For Distillation", *Water Conditioning and Purification*, 32(10):46-52, (November, 1990).

57. McCabe, L. J. and Winton, E. F. "Studies Relating to Water Mineralization and Health", *Journal American Water Works Association*, 62(1):26-30, (January, 1990).

58. Hanneman, R. L., "Evidence Challenges Sodium-Health Connection", *Water Technology*, 12(10):30-31, (October 1989).

59. McGarvey, F. X. and Ziarkowski, S. M., "Choosing the Right Resin", *Water Technology*, 13(9):29-33, (September, 1990).

60. Nebel, C. "Ozone", Kirk-Othmer Encyclopedia of Chemical Technology, New York, NY, Wiley Interscience, 698, (1981).

61. Ireland, J. C., In-House Disinfection/Disinfection By-Product Summary, Drinking Water Research Division, Cincinnati, Ohio, (1990).

INTRODUCTION

The case studies presented in this chapter describe the events, issues, and challenges that confront communities and individual homeowners when their drinking water is found to be contaminated. These studies describe the economic, bureaucratic, technological, and institutional lessons learned while trying to produce potable drinking water.

SUFFOLK COUNTY, NY

Background

The groundwater supply for Suffolk County, NY, (most of Long Island) has been designated as a sole-source aquifer under the Safe Drinking Water Act. In recent years, concern has developed about the contamination of this groundwater by agricultural chemicals (fertilizers, insecticides, herbicides, nematocides, fungicides). This concern increased when specific chemicals were identified in private drinking wells.

Since 1978, Suffolk County has examined groundwater for agricultural and organic contaminants as well as for their decay products. During this testing, 101 agricultural or organic compounds were evaluated; 41 were found in the groundwater. Many of these contaminants were present in trace quantities (1-10 μg/L), but four agricultural compounds were found to be present at levels > 100 μg/L -- aldicarb; carbofuran; 1,2-dichloropropane (1,2-DCP); and 1,2,3-trichloropropane (1,2,3-TCP). Nitrates from fertilizer applications were also present in quantities exceeding the primary drinking water standard (up to 15 mg/L).[1]

Agriculture has been a major industry in Suffolk County for over 200 years. The potato plant (a principal crop) is susceptible to a number of pests, most notably the golden nematode, which attacks the roots, and the Colorado potato beetle, which eats the leaves. Since the early 1950's, pesticides containing 1,2 dichloropropane have been applied to fields infested with golden nematodes, particularly those fields quarantined by the U.S. Department of Agriculture. In 1974, the carbamate pesticide aldicarb (trademark TEMIK, Union Carbide Corp.) was registered for use on potatoes, and by 1976 the chemical was being used by all growers at an application rate of 3 pounds of active aldicarb per acre.[2]

Aldicarb was used extensively for four growing seasons in Suffolk County until in August 1979, several drinking water supply wells in the County near potato farming operations were found to be contaminated by aldicarb. An initial testing program detected the presence of aldicarb residues in drinking water supply wells at 61 homes adjacent to potato farms. Aldicarb had been used on more than 22,000 acres of potato crops in Suffolk County. As a result, a major monitoring program was undertaken to determine the extent of the contamination. During an eight-week period between April and June of 1980, more than 8,000 water samples were analyzed. The results of the testing indicated approximately 15% of the wells had aldicarb residues above a level considered safe for drinking purposes.

The New York State Health Department established a 7 part per billion (ppb) allowable guideline level for aldicarb residues in drinking water. Based upon the guideline level, the Suffolk County Department of Health Services advised residents having aldicarb residues in excess of the guideline not to use the water for consumptive purposes.

The parent compound (aldicarb) was not detected in Suffolk County groundwater but the aldicarb metabolites (aldicarb sulfone and aldicarb sulfoxide) occurred as follows:

total aldicarb = parent aldicarb + aldicarb sulfone + aldicarb sulfoxide
  (100%)   (0%)   (40-60%)   (40-60%)

Some representative data from community private wells are shown in Table 1 to illustrate the contamination problem.

TABLE 1.   REPRESENTATIVE ALDICARB CONCENTRATIONS - SUFFOLK COUNTY GROUNDWATER

Community	Wells Sampled	>7 ppb	1-7 ppb	% Below Detection
1	222	2	18	91.0
2	434	43	46	79.5
3	2,161	351	345	68.8
4	1,832	270	256	71.3
5	3,160	359	374	76.8

ppb = μg/L

1,2 DCP testing began in 1980, and only a few agricultural communities were found to be contaminated. It is suspected that the primary sources of this chemical are several pesticides (DD, Vidden D, Vorlex Telone)--fumigants used for golden nematode control--each containing DCP. The chemical is no longer used by the Department of Agriculture as a fumigant on Long Island.

In one community, DCP was found in 17 of 33 wells, with two wells approaching or exceeding the State Health guidelines of 50 μg/L. A second community had 2 of 9 samples contaminated at levels of 10-15 μg/L, and a third area had a private well with a concentration of 49 μg/L.[3]

Aldicarb's use was discontinued when the manufacturer, Union Carbide, first discovered the chemical in Long Island groundwater. Union Carbide requested and received approval to modify their labeling permit to prohibit sale in Suffolk County. Although aldicarb was no longer used in Suffolk County, Cornell University and Suffolk County estimated that over 100 years would be required to purge the aquifer of aldicarb based on the flow pattern of the groundwater.[3]

Carbofuran was available for agricultural use before and after aldicarb. The amounts found in groundwater have been much less in number of wells and concentration. As an example, county records show only 1.8 percent of 2,000 wells sampled in 1985 exceeded state guidelines, compared to 11.7 percent exceeded for aldicarb.

Fertilizer practices, with as much as 114 Kg N/acre applied, led to widespread nitrate contamination of the shallow aquifer.[4] The many years of fertilizing has taken its toll on groundwater quality. The severity of the problem is reflected in Table 2 by concentrations of nitrate found in private wells from two communities in Suffolk County.

TABLE 2. NITRATE CONCENTRATIONS IN HOUSEHOLD WELLS

Community	Wells Sampled	(NO_3 -N)		
		0-5 mg/L	5-10 mg/L	10 mg/L
1	635	372	163	100
2	1,121	575	354	192

Once the nitrate passes through the topsoil (15 to 30 cm avg.), the coarse sand and gravel of the upper geological formation allow the nitrate to percolate unchanged directly to the groundwater table. Local hydrology is such that aquifer segments have been affected.

GAC Evaluations

Five treatment units were pilot tested to determine removal efficiencies using actual Long Island groundwater (Table 3). The two POU units treated less than 1,000 gallons of water before exceeding the New York State guideline of 7 μg/L. The three POE units did prove to be effective and the decision was made to offer affected homeowners an equivalent GAC system.

TABLE 3. DESCRIPTION OF TREATMENT UNITS

Model and Type Unit	Filter Media	Size of Carbon Column	Amount of Carbon
POE	12/40 Mesh	9" dia. X 20"	17-1/2 lbs.
POE	20/40 Mesh	9" dia. X 20"	15 lbs.
POE	12/40 Mesh	9" dia. X 20"	17-1/2 lbs.
POU	18/40 Mesh	3" dia X 8-1/2"	1.04 lbs.
POU	12/40 Mesh	3" dia X 8-1/2"	0.937 lbs.

Union Carbide Corporation agreed in June 1980 to provide, at no cost to the affected residents, granular activated carbon (GAC) treatment systems for all water sources exceeding the New York State guideline level. From the viewpoint of the County, the GAC treatment systems were to be an immediate response to an emergency situation and were not to be considered a long term final solution.

Over 1,000 private wells had been found to contain aldicarb above guideline levels by the completion of the intensive sampling program in June 1980. Since it would not be possible to maintain a continuing testing program to determine the effectiveness of each individual GAC treatment unit, representative filter locations were selected for testing.

The effectiveness of the POE units as a percentage of actual theoretical life ranged from a low of 37% to a high of 158%. Over 100 POE units were eventually monitored on a bi-monthly basis with 93% operating satisfactorily as designed.

Operation and Maintenance Problems

Investigation of treatment units where premature breakthrough of aldicarb residues occurred revealed several instances of installation or operation and maintenance difficulties. Common problems included:

-    Failure to have treatment unit in automatic backwash mode or not having the unit connected to electrical outlet.

-    Raw untreated water bypassing filter resulting from piping arrangements which cannot be easily determined; i.e., buried pipe in concrete slabs or concealed piping in walls.

-    Failure of homeowners to place filter back into treatment mode after manually bypassing system for lawn watering, etc.

-    Inadequate backwashing cycle resulting in plugging or reduction of water pressure through the filter caused by accumulation of sediment.

-    Mechanical failure of some components of the treatment unit and plumbing accessories caused by sediment blockage and/or possible corrosive breakdown of treatment unit materials.

Microbiological Concerns

The possible impact of microbial activity in the carbon media was investigated as part of the bi-monthly monitoring program. Analyses were performed for total coliform, standard plate count and *pseudomonas*.

Concern had been raised regarding the possibility of bacterial organisms accumulating and growing in the carbon media. Test results indicated microbial activity was not a problem.

One of the reasons for the lack of bacterial problems was the general absence of bacteria in Long Island groundwaters. Past experience with groundwater conditions reveal bacteria are effectively removed by the soils. Since an originating source of bacteria is not present, the possibility of accumulation and growth of bacteria in the carbon media is remote.

In addition, the GAC treatment systems were provided with backwash capability which would tend to mitigate the possibility of bacterial growth. The backwash process not only removed any sediment, such as iron that is present, but also flushed the media of any potential bacterial growth.

Although recommendations have been made by some agencies to provide disinfection for GAC treatment systems used in some communities, there is no apparent need for disinfection under Long Island conditions.

Used Carbon Disposal

Arrangements for proper disposal of used carbon from recharging operations in an environmentally safe manner were made by Union Carbide Corporation. The used carbon was shipped out of state to a carbon manufacturer where it was regenerated via a high temperature (1,800 degrees F) process. The regenerated carbon was recycled for use in industrial applications. Only virgin carbon was used in the treatment units being provided under the filter program.

Further Suffolk County Studies

Because of the seemingly random or patchwork occurrences of nitrate and various agricultural chemical contaminations of Long Island groundwater, two other studies evaluating RO and IEX were undertaken. POU/POE systems must be flexible and robust in their design to accommodate such variety and ensure consumer safety.

A POU/POE demonstration project in the towns of Riverhead and Southhold located in the northeastern portion (North Folk) of Long Island was started in September 1985 and completed in mid-1986. Future land use projections by the regional planning board indicated that low density residential development would significantly increase in both towns, whereas agricultural usage would decline by approximately 18 percent. Thus, with the increased residential development, the demand for potable water for domestic use would continue to grow.

The two towns rely exclusively on groundwater for water supply. Because of the rural nature of the communities, public water systems are limited and individual private wells are used to supply the majority of homes.

Groundwater quality problems on the North Fork were severe. Shallow, thin groundwater aquifers were extensively contaminated by agricultural chemicals (nitrates, pesticides, and herbicides) and were threatened by saltwater intrusion from overpumping.

In 1983, a study evaluated different water supply options for the various communities and found that, because of the rural nature of many of the communities, the provision of public water supply systems throughout the contaminated areas would be prohibitively expensive. Individual home water treatment systems were recommended for many rural areas where the groundwater was contaminated.[5]

A total of 10 manufacturers participated; 18 units were installed in homes receiving water with varying types of contamination. The performance of the units was monitored through sampling and analysis of the raw and treated waters.

Table 4 summarizes the water quality problems and the types of devices installed. Table 5 presents a summary of the performance of each of the 18 units.

All units demonstrated satisfactorily their ability to remove the contaminants of concern, and the consumers were satisfied with their units. Because the reliability of operation varied between units, longer operating histories and records would be necessary to prove the efficiency of the devices. Given the supervised maintenance program, however, the units were shown to be viable.

Sampling results uncovered some problems that could be traced back to inadequate installation or maintenance or both. They include the following:

- Positive coliform counts were detected in the effluent. Three different units had coliforms in initial samples after installation; resamples proved satisfactory. Two of these units were countertop models, which would be most susceptible to cross contamination by the homeowner. These units also contained silver-impregnated carbon, which is supposed to be bactericidal. More positive disinfection procedures should be followed before and after installation of these devices.

TABLE 4.   POINT-OF-USE TREATMENT UNITS INSTALLED AT
SUFFOLK, COUNTY, NEW YORK

Unit Number	Water Quality Problem	Type of Device (Treatment Method)
1	Nitrate	Countertop (carbon + ion exchange)
2	Nitrate	Countertop (carbon + ion exchange)
3	Nitrate, chloride	Third tap (RO + carbon)
4	Nitrate	Third tap (RO + carbon)
5	Nitrate, chloride	Third tap (RO + carbon)
6	Nitrate, VOC	Countertop (distiller)
7	Nitrate	Third tap (RO + carbon)
8	Nitrate	Third tap (RO + carbon)
9	VOC	Whole house (carbon)
10	Nitrate	Third tap (RO + carbon)
11	VOC	Whole house (carbon)
12	Nitrate	Batch (distiller)
13	Iron, carbofuran	Countertop (filter + carbon)
14	Manganese	Third tap (RO + carbon)
15	Nitrate	Third tap (RO + carbon)
16	Iron	Whole house (aeration + carbon + filter)
17	Nitrate	Third tap (RO + carbon)
18	Nitrate	Whole house (ion exchange)

TABLE 5.   SUMMARY OF PERFORMANCE OF POINT-OF-USE DEVICES
(SUFFOLK COUNTY, NEW YORK)

Unit Number	Average Nitrate -- mg/L		Average Organics -- μg/L	
	Raw	Treated	Raw	Treated
1	9.2	3.3		
2	7.7	2.4		
3	10.8	4.6		
4	9.9	4.3		
5	0.4	<0.2		
6	12.2	<0.2	12	<2
7	11.1	0.3		
8	7.7	0.2		
9			58	<2
10	11.2	0.3		
11			53	<2
12	9.3	0.2		
13	2.3*	0.1*	13+	<2+
14	1.7#	0.07#		
15	8.6	0.8		
16	0.96*	0.1*		
17	11.5	0.3		
18	12.1	0.6		

*   Average values for iron   + Average values for carbofuran
\#   Average values for manganese

- Plate counts were present in samples from the carbon units. Fifteen devices used carbon as either the primary treatment medium or for secondary treatment. Thirteen raw and treated samples were taken from nine devices (Table 6). The raw water data showed plate counts in 10 of 13 samples, with a range of 1 to 1,100/mL. The treated water showed counts in 11 of 13 samples, with a range of 4 to >6,000/mL. In one instance, the treated water count was less than the raw water count. Three whole-house units (9, 11, and 18) showed results that were quite different from each other, yet similar in range to those of the six other devices. The data confirm the observations of other researchers that bacteria will grow on granular activated carbon. However, no evidence of pathogenic bacteria growth has been isolated, even with elevated plate counts.

- Three units were found to have organics in the effluent that were not found in the raw water. In one instance, a plastic collector provided with the unit was the source. In the other two cases, leaching of solvents used in product manufacture or assembly may have been the cause.

Several units developed plumbing leaks that required repair. Two units had copper discharge lines that caused high levels of copper in the effluent. The lines were replaced with standard plastic tubing. After the first water sampling, one unit was found to be defective, and it was replaced.

Reverse osmosis units 3 and 4 had much lower efficiencies than units 7 and 8, probably because of the lower efficiency of cellulose acetate membranes compared with thin-film composites. One whole-house carbon unit (11) was exhausted after five and half months of operation. By comparison, unit 9, which contained twice as much carbon (1.0 cu ft [0.028 m^3]), operated for almost a year. Although this comparison ignores flow and concentration (unit 9 had higher raw water concentrations of 1,1,1-trichloroethane), it indicates a need to have either two 0.5 cu ft (0.014 m^3) cartridges in series or one 1.0 cu ft (0.028 m^3) container with a six-month-plus operating life.

Cost Estimates

Table 7 includes representative costs for the various types of equipment evaluated in this study. Table 8 presents total average annual costs per home and includes capitalized first costs, annual operation and maintenance costs, and annual costs for monitoring and administering a point-of-use program. Initial costs were estimated from manufacturers' literature, as were annual operation and maintenance costs. If units were purchased in bulk, the cost would be lower. The cost ranges represent units of different capacities.

TABLE 6.   PLATE COUNT DATA FOR POINT-OF-USE DEVICES
(SUFFOLK COUNTY, NEW YORK)

Unit Number	Plate Count in Raw Water total plate count/mL	Plate Count in Treated Water total plate count/mL
5	37	2,100
7	10	1,300
7	<1	220
8	6	>6,000
8	4	>6,000
9	1,100	1,600
11	<1	4
12*	<1	<1
12	1	<1
15	6	2,100
17	18	4
17	100	>6,000
18	25	36

*   Distillation unit

TABLE 7.   REPRESENTATIVE COST RANGES FOR POU/POE TREATMENT UNITS
(SUFFOLK COUNTY, NEW YORK)

Unit Type	Single Tap		Whole Tap	
	Initial Cost $	Installation Cost $	Initial Cost $	Installation Cost $
Reverse osmosis	500-800	70-150	6,000-8,500*	250-350
Granular activated carbon	200-350	60-100	1,100- 3,000	75-150
Ion exchange	100-300	60-100	1,500- 2,000	150-200
Membrane filtration	150-200	80-100	1,500- 2,000	150-200
Distillation	200-800	100-150	9,500-11,000	200-300

*   This includes the RO unit, the storage tank, and a dispenser pump.

TABLE 8.   AVERAGE ANNUAL COSTS PER HOME OF POINT-OF-USE
TREATMENT DEVICES

COST CATEGORY	COST
Amortized equipment cost	$160-170
Operation and maintenance	50-100
Monitoring and administrative costs	15-20
Total	$225-290

Post-Script to Long Island Experience

In 1985, there were about 2,500 private homes using GAC POE units to remove primarily aldicarb from their drinking water. In 1990, that number leveled off to 3,400.[6] Suffolk County's largest community supplier installed GAC at 58 wells costing approximately $30 million. Nitrates and other organic contaminants are still being found in Long Island groundwater.

For example, Dacthal (DCPA) has recently been found in 15 wells exceeding state drinking water standards. Dacthal -- a trade name for the chemical dimethyl tetrachloroterephthalate -- has been used in Suffolk County since the 1950's to control weeds on lawns, sod farms and on crops including onions, cauliflower, strawberries, and tomatoes.

An estimated 10 tons of the herbicide was applied in Suffolk County in 1982, the last year for which an estimate is available. The amount probably has remained fairly constant.

Beginning in 1982, county officials, with the company's cooperation, began sampling local wells for traces of the chemical. Of the 198 wells tested over five years, 43 were found to contain traces of a byproduct of Dacthal, and 15 of those contained concentrations exceeding the 50 parts-per-billion limit established by the state for all organic chemicals, regardless of their toxicity.

Although some evidence of contamination has been available to the county since 1979, officials delayed taking action against the company in the hopes of arriving at an amicable settlement similar to that reached with the maker of Temik. The specific health effects of Dacthal are not known, although it has been shown to cause liver disfunction in laboratory animals.

Dacthal is one of about 13 agricultural chemicals that have been detected in drinking water in Suffolk County. State and county officials said they targeted Dacthal because of its widespread use and high concentrations found in the water, up to 20 times the legal limit.

Traces of the chemical have been found in public wells, but because the county does not have the ability to perform routine tests for Dacthal, the extent of contamination in the public water supply is not known. This legacy of legal, routine application of agricultural chemicals highlights the need to be conservative in the design of POU/ POE treatment devices because of competitive effects from contaminants possibly yet to be discovered in one's drinking water supply.[7]

Another legacy of Long Island's saga has been the deluge of literature that used a combination of misinformation, distortion, and lies to scare and trick homeowners into purchasing home treatment systems. A small percent of the POU/POE industry used the skull and crossbones, improperly marketed EPA registration, excerpted statements out of context, and flashed WQA approvals and logos to sell their products. As a result, there is a host of local (Suffolk County) and State (New York) laws proposed to regulate the industry.[8]

CASE STUDIES   207

ELKHART, INDIANA

Background

To date, five separate Superfund actions have been taken in and
around Elkhart, Indiana. The city of Elkhart is located in north central
Indiana at the confluence of the Saint Joseph and Elkhart Rivers. The
population of Elkhart is approximately 40,000, while the areas adjacent
to the city host an additional 25,000 people. Elkhart is a diversified
industrial community manufacturing pharmaceuticals, band instruments,
recreational vehicles, and injection molded plastics and foams. A large
number of industries in the Elkhart area use or have used TCE or other
organic solvents in their processes.

In the fall of 1984 a private citizen sampled his well and found
trichloroethylene (TCE) in excess of 200 μg/L. The citizen contacted
neighbors and advised the Elkhart County Health Department of the situation. The County Health Department initiated a sampling program and
complied a list of residential wells in the area. The County and the
State collected several dozen samples. Analyses indicated that the
problem was more extensive than anticipated, and the United States
Environmental Protection Agency (USEPA) was contacted.

It was found that private wells in the city of Elkhart were contaminated by several volatile organic compounds. These included trichloroethylene (TCE), dichloroethylene (DCE), perchloroethylene (PCE),
trichloroethane (TCA), and carbon tetrachloride (CCL_4). The levels of
contamination ranged up to 19,380 μg/L of TCE. Drinking water from these
contaminated wells constituted an immediate and significant threat to the
residents of the affected households. In addition, a real threat is
present from the inhalation and absorption of water contaminated with TCE
in levels above 1,500 μg/L according to the Agency for Toxic Substance and
Disease Registry (ATSDR).

In May of 1985, EPA initiated a city-wide sampling program in which
over 500 samples were analyzed. Over 80 wells were contaminated in excess
of 200 μg/L in the eastern part of Elkhart. In addition, 15 of the wells
were in excess of 1,500 μg/L. Carbon tetrachloride was also found in this
area. As a result, over 800 residents were placed on bottled water
delivery. It was decided to extend city water mains to those affected
areas because of the widespread contamination.

In total, approximately 14,500 feet of water main were installed
ranging in size from 1-1/2 inch to 12 inch, and 301 homes and 7 businesses
were connected to the municipal system. In addition, 11 homes with minor
contamination not adjacent to a water main were given "point-of-use" water
filters. Two homeowners refused to have city water hooked up to their
homes despite repeated efforts to convince them of the threats posed by
the VOC contamination.

This however, was not the limit of Elkhart's experience with POU/ POE water treatment devices. In June of 1986, once again, a private citizen, on the west side of town had his well water analyzed. The results indicated severe contamination with both TCE and carbon tetra-chloride. Levels of 800 μg/L and 488 μg/L respectively, were found. EPA sampled eighty-eight private wells and found carbon tetrachloride values as high as 6,860 μg/L and TCE values of 4,870 μg/L. EPA then decided to install GAC POE units because of the time and distances involved in attempting to extend the city's water mains. Fifty-four POE filters were installed and twenty-two POU filters were selected for homes with slight contamination near areas of very high concentrations as a safeguard if and when the plume expanded.[9]

Two of the homes were equipped with packed tower air strippers with two GAC POE units in-series. This unique configuration is displayed in Figure 1. Photographs of one of the units installed under the basement steps at one of the homes is shown in Chapter 3, Figure 18. The air stripper has a 40:1 air-to-water ratio and operates at a rate of 5 gpm. The air stripper is packed with 1 inch diameter polypropylene cylinders. The entire unit installed cost is about $4,000. There have not been any microbiological problems encountered to-date although flushing has been recommended when the unit remains idle for more than one day. The unit is capable of having a UV light for installation of post-GAC disinfection.

POE Performance

Granular activated carbon isotherm calculations have proven unreli-able in predicting breakthrough for the GAC POE units. Significant under and over estimating of time to breakthrough has been encountered. Com-petitive effects are very evident in a dual GAC unit monitored for a special EPA study in Elkhart. Isotherm data estimated breakthrough for chloroform at approximately 225,000 gallons but it was estimated to actually have broken through at approximately 130,000 gallons. The following data indicate the risk for homeowners if special monitoring had not been undertaken and carbon change-out was only scheduled on a routine basis (Table 9).

TABLE 9.   ELKHART, INDIANA GAC BREAKTHROUGH DATA

Treated Volume (gallons)	Contaminant	Influent Concentration (μg/L)	Effluent Concentration (μg/L)
173,990	Chloroform	373	30
	Carbon tet	≈4,400	112
	TCE	162	< 1
188,210	Chloroform	356	58
	Carbon tet	≈4,300	224
	TCE	138	< 1
198,370	Chloroform	308	36
	Carbon tet	≈3,900	309
	TCE	111	< 1

   The number of gallons treated before breakthrough has ranged from 25,000 to over 300,000 gallons. Figure 1 displays the results of one home with very early breakthrough. As can be seen, there was very low water usage over time and significant levels of carbon tetrachloride and TCE that could be exhibiting competitive effects, thus the early breakthrough. Table 10 summarizes GAC breakthrough for the major organics found in Elkhart's well waters at homes that have experienced breakthrough and are possibly more illustrative of GAC capabilities. Figures 2, 3, and 4 show removal of carbon tetrachloride, chloroform, and 1,1,1-trichloroethane respectively for a residence using two carbon columns (13" dia X 50" carbon depth), with each unit containing 110 pounds of Calgon FS-300, 20 X 50 mesh size GAC. An average methylene chloride contamination level of 115 μg/L was consistently removed to below detection levels throughout the study period, as well. Carbon replacement cost in 1989 dollars was approximately $510 per tank and $40 each for the sediment pre-filter. There are currently several drums of spent carbon awaiting a decision on how they are to be disposed.

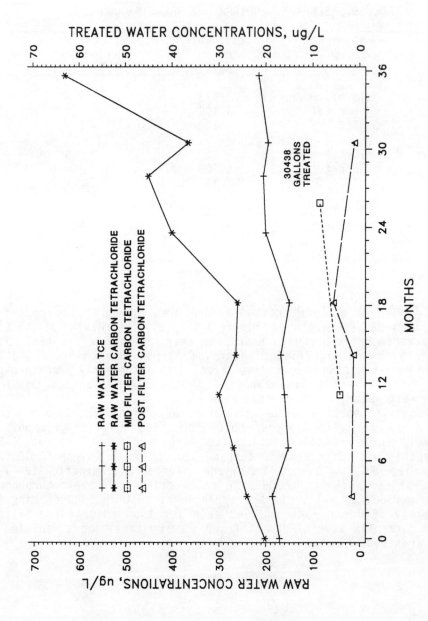

FIGURE 1. ELKHART, INDIANA GAC POE PERFORMANCE

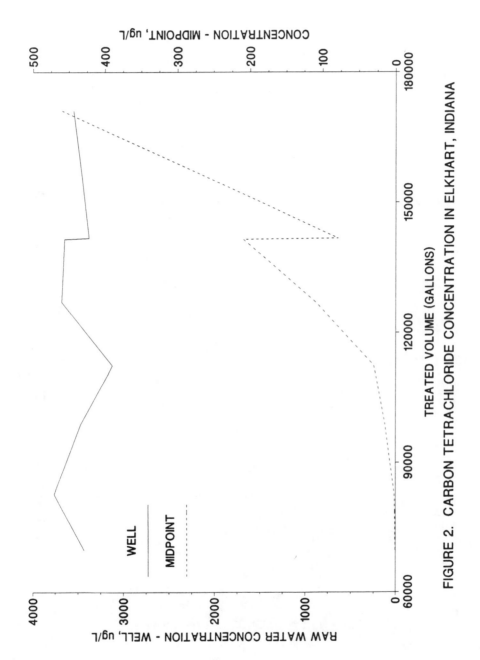

FIGURE 2. CARBON TETRACHLORIDE CONCENTRATION IN ELKHART, INDIANA

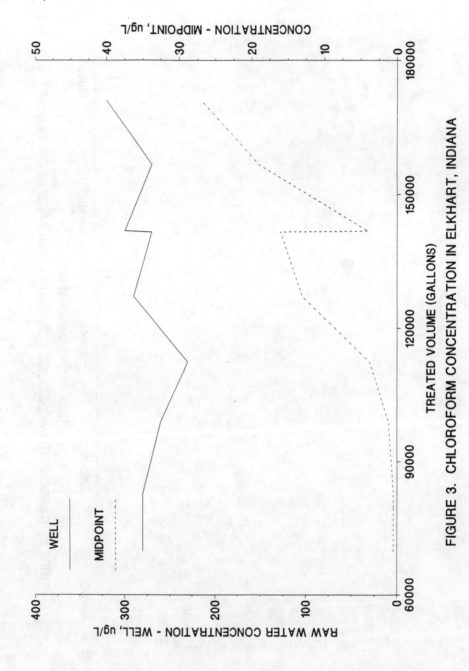

FIGURE 3.   CHLOROFORM CONCENTRATION IN ELKHART, INDIANA

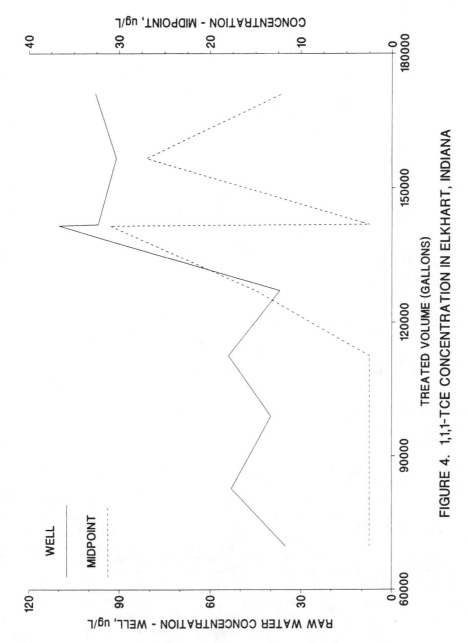

FIGURE 4. 1,1,1-TCE CONCENTRATION IN ELKHART, INDIANA

TABLE 10. ELKHART, INDIANA GAC BREAKTHROUGH SUMMARY FOR
CARBON TETRACHLORIDE

SITE	AVG. INFLUENT CONCENTRATIONS (μg/L)			GALLONS TREATED	MONTHS	POSSIBLE BREAKTHROUGH CAUSES
	TCE	CCL_4	$CHCL_3$			
1	170	291	15	30,500	25	Competitive effects, bacterial colonization
2	60	2,864	ND	120,000	22	High influent levels
3	418	2,188	ND	150,000	24	High influent levels
4	331	135	10	135,000	16	Competitive effects, TCE
5	1,686	348	50	140,000	18	TCE Breakthrough

ND = Not detected

FRESNO, CALIFORNIA

Background

This case study is based on a 1988 study between the EPA and California State University, Fresno that highlights the many issues surrounding the viability of POE treatment devices.[10] The provision of safe water to residents in areas of low population density is an ever increasing problem in California's San Joaquin Valley. A survey indicated that 99 of 231 test wells (42.9%) were found positive for organic contamination in Fresno County. Los Angeles County was second in the state with 41.2% testing positive. Hundreds of GAC units have already been installed in the Fresno, California area alone by local water conditioning firms on private water wells contaminated with dibromochloropropane (DBCP). The units being marketed have been approved in concept by the Sanitary Engineering Branch, California Department of Health Services, Fresno, California. Typical domestic water well production values are 200 to 1,000 gallons per day. The GAC units are equipped with a flow totalizer, a flow restriction to maintain a minimum detention time (1.5 minutes), pressure gauges at the inlet and outlet of the unit, and facilities to backwash the carbon to control head losses. These units are sold or leased to the user and serviced by local water conditioning firms who are to retrieve GAC units when they require replacement. Ten of these units supplied to private individuals in the rural south-east Fresno area by a POE treatment unit dealership were selected for study, and their feed and product waters were sampled for DBCP on a four to eight week basis over two years.

Results

The ten sites studied provided a fairly complete picture of what homeowners can expect from POE devices. Sites 1 and 2 showed continuous removal of DBCP from the feed water to or below the detection limit of 0.01 μg/L (EPA proposed MCL = 0.02 μg/L). Figures 5 and 6 display the results of this study. The feed water at Sites 1 and 2 contained relatively low concentrations of DBCP. However, the DBCP concentration at Site 2 appeared to increase during the period of the study. This increase in DBCP concentration decreased the amount of water that the unit could process before the GAC was exhausted. This points up the need for continual surveillance of GAC units.

Initial results for the Site 3 POE unit (Figure 7) indicated DBCP concentrations in the product water comparable to those in the feed water. However, by May 1986 the product water DBCP concentration increased to 1.10 μg/L, twice the feed water DBCP concentration of 0.54 μg/L. The product water DBCP concentration at Site 3 has consistently been greater than the feed water concentration. The retired owners of this unit were aware of its poor performance, but they indicated that the cost of replacing the GAC was too great for them.

At the initiation of the study in November 1985, the domestic water supply used by the household at Site 4 (Figure 8) had a DBCP concentration of 3.64 μg/L. The product water DBCP concentration remained high until early 1986 when the discovery was made that erroneous plumbing connections had been made by a plumber (working on the owner's swimming pool). He connected the house plumbing to another well which was not connected to the GAC POE unit. While no statement can be made as to the frequency of problems of this nature caused by improper plumbing, this problem was detected because of the sampling which was carried out as part of this study. Similar problems existing with POE units owned by other individuals might also be discovered if routine monitoring of the units was made a general requirement for their use.

The GAC POE units at Sites 5, 6, and 7 performed well throughout the study (Figures 9, 10, and 11, respectively). The DBCP concentration in the product water varied from not detectable to 0.10 μg/L.

The GAC in the Site 8 POE unit (Figure 12) was nearing exhaustion since the product water's DBCP concentration was generally increasing from March 1986. The greatly improved performance (0.04 μg/L) for the Site 8 POE unit evidenced for March 1988 was the result of the replacement of the POE unit's GAC. The owners of this unit, as most GAC POE unit owners,

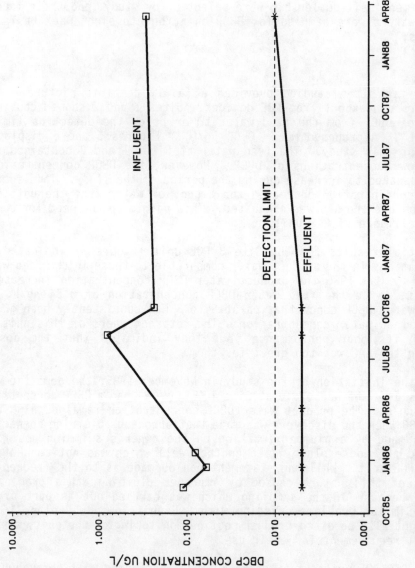

FIGURE 5. SITE 1 GAC POE PERFORMANCE IN FRESNO, CALIFORNIA

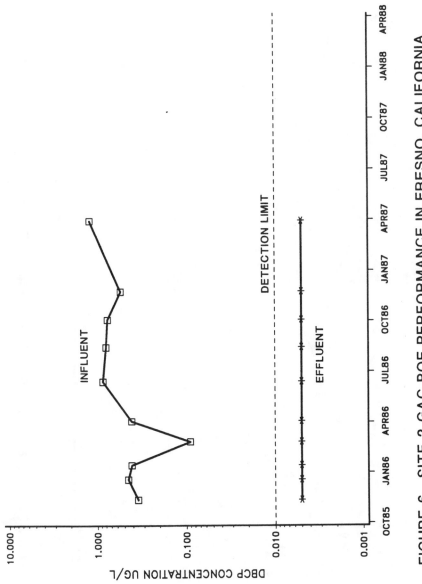

FIGURE 6.   SITE 2 GAC POE PERFORMANCE IN FRESNO, CALIFORNIA

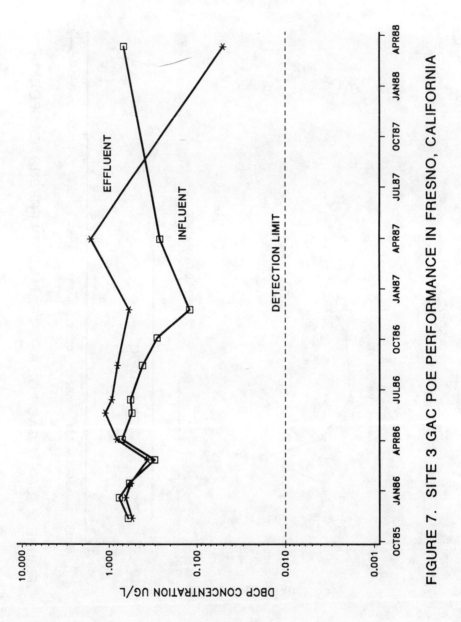

FIGURE 7. SITE 3 GAC POE PERFORMANCE IN FRESNO, CALIFORNIA

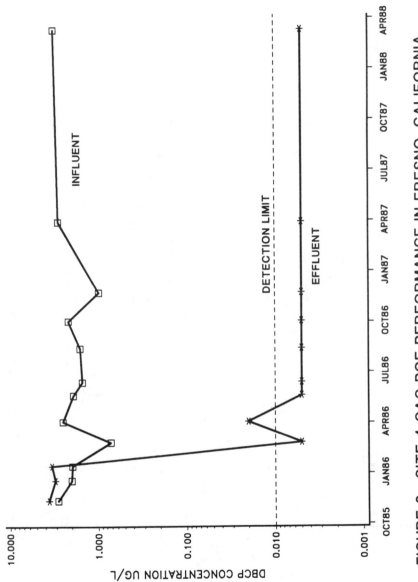

FIGURE 8. SITE 4 GAC POE PERFORMANCE IN FRESNO, CALIFORNIA

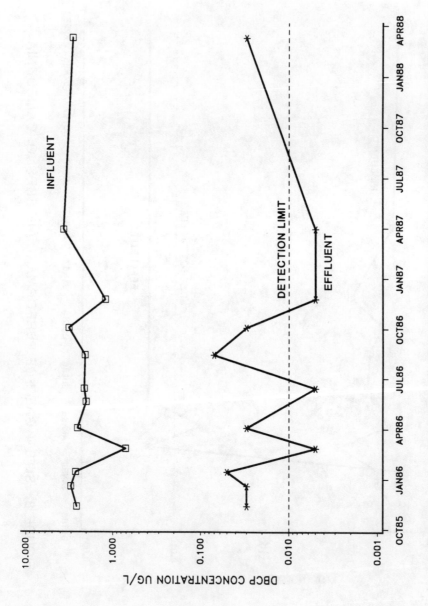

FIGURE 9   SITE 5 GAC POE PERFORMANCE IN FRESNO, CALIFORNIA

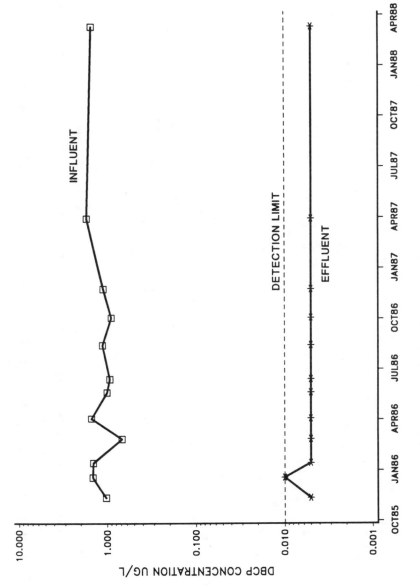

FIGURE 10. SITE 6 GAC POE PERFORMANCE IN FRESNO, CALIFORNIA

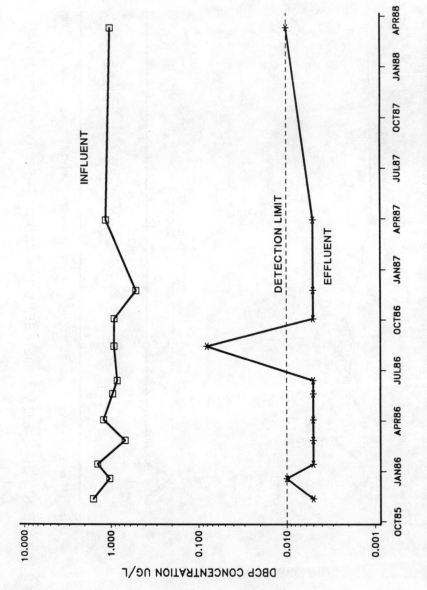

FIGURE 11. SITE 7 GAC POE PERFORMANCE IN FRESNO, CALIFORNIA

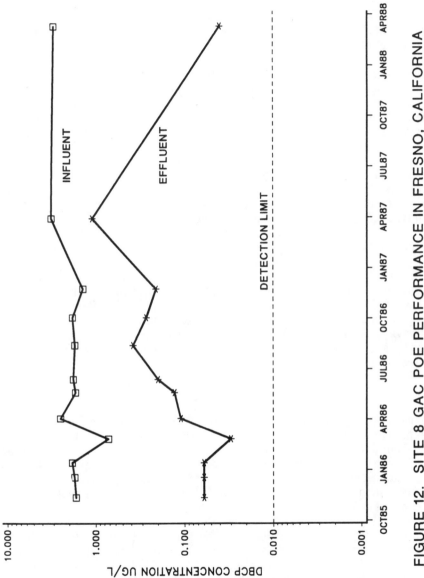

FIGURE 12. SITE 8 GAC POE PERFORMANCE IN FRESNO, CALIFORNIA

had no means of being routinely informed of the DBCP concentration in either their product water or feed water without purchasing expensive sampling and analytical services. And they must also initiate the actions necessary to obtain sampling and analysis.

The Site 9 data (Figure 13) appears to indicate that the POE unit's GAC was nearing exhaustion (DBCP values consistently near or above 0.1 μg/L, and generally increasing with time). This unit was installed by the previous homeowners at this site. The home has since been sold and the homeowners at the end of this study owned a GAC POE unit which was providing limited protection.

The Site 10 data (Figure 14) showed initial high product water DBCP concentrations. When informed of these results, the owners had the GAC replaced. Subsequent data showed mostly undetectable DBCP concentrations resulting from a change of the unit's GAC.

Microbiological Concerns

Heterotrophic plate counts (HPC) were made on seven sets of water samples collected from the feed water and product water of the ten monitored POE units from February 1986 through March 1987. Results were scattered and inconclusive regarding bacteriological growth in the carbon beds. However, an inspection of the data seems to suggest that the number of HPC organisms generally increased in the product waters relative to the feed waters.

Conclusions

The results obtained from monitoring the ten GAC POE devices showed that the performance of these units changed markedly over short periods of time. Thus, these units required conscientious, periodic monitoring. This unit operation requirement was not carried out by either the owners or the vendor; the owners generally lacked the expertise and, in many cases, the vendor had no contractual authority or responsibility to monitor the POE units.

Monitoring of the operation of POE (and POU) units appears to be a significant shortcoming in the application of this technology in most areas in the United States. During the course of this study, operation and maintenance problems were encountered with three of the ten units monitored, and these problems were significantly severe in two of the three cases where water delivered to consumers had DBCP concentrations exceeding the 0.02 μg/L proposed MCL.

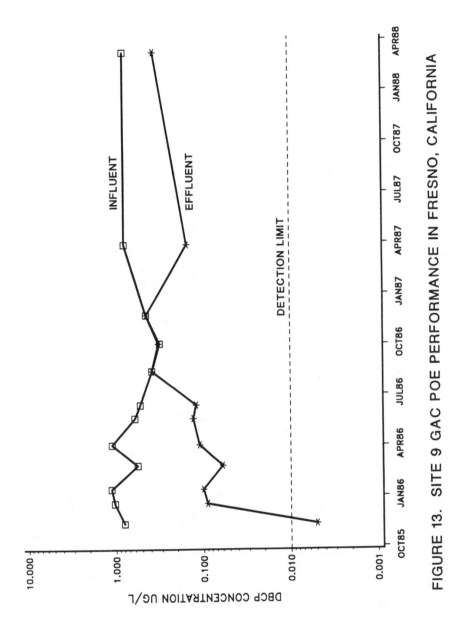

FIGURE 13. SITE 9 GAC POE PERFORMANCE IN FRESNO, CALIFORNIA

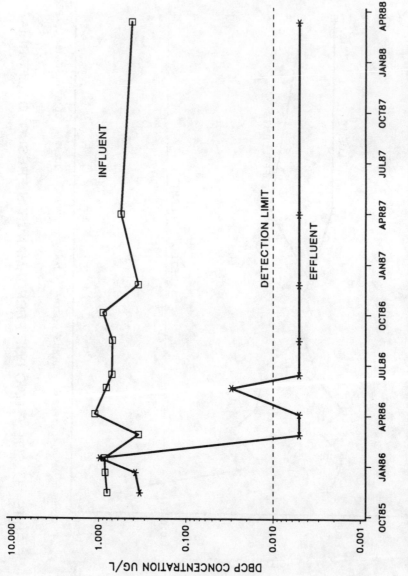

FIGURE 14.   SITE 10 GAC POE PERFORMANCE IN FRESNO, CALIFORNIA

SAN YSIDRO, NEW MEXICO

Background

An EPA study utilized POU treatment for removal of arsenic and fluoride from the community of San Ysidro, New Mexico.[11] The Village of San Ysidro is a small rural community of approximately 200 people located in the north central part of New Mexico approximately 45 miles north of Albuquerque.

The Village has had a long history of water supply problems including low water pressure, no water at all, and quality problems including taste, color, clarity and odor in addition to arsenic and fluoride contamination and sporadic coliform violations.

The Village water supply source is an infiltration gallery which produces an average of 27,000 gpd in winter and 36,000 gpd in summer from the groundwater. The infiltration gallery has a storage capacity of 17,000 gallons. The Village currently uses an average of 30,000 gpd, which equates to about 150 gpd per person. This consumption rate pushes the production/storage capacity limits of the gallery.

A 20,000 gallon elevated storage tank connected into the distribution system should provide the additional capacity the Village needs, but there have been problems in keeping the tank operational. First, the pumps which were located in the infiltration gallery did not have adequate controls to allow them to operate automatically to maintain an adequate supply of water in the tank. There was no remote readout on the status of the pumps or system. The only way to know there was a problem was when a faucet was opened and little or no water was delivered. Secondly, the Village did not have an assigned person who knew the system and who had the responsibility to keep it operating. A Village employee, the Mayor or one of the Village council, usually turned the pumps on when someone called to complain about the low pressure or to report that they have no water. The pumps should run all night to fill the tank, but because of other problems with the controls and overheating pump motors, someone would need to monitor them all night to ensure safe operation.

The local groundwater contains leachate from geothermal activity in the area's abundant mineral deposits and is therefore high in mineral content. The groundwater exceeded the recommended standards and/or MCLs for arsenic, fluoride, iron, manganese, chloride and total dissolved solids. The contaminants of concern in the Village water supply were arsenic V and III and fluoride which exceeded the MCLs (Table 11) by three to four times.

TABLE 11.   MAXIMUM CONTAMINANT CONCENTRATION VS. MCL FOR
SAN YSIDRO, NEW MEXICO

Contaminant	Maximum Contaminant Concentration (mg/L)	Maximum Contaminant Level or Recommended Standard (mg/L)
Iron	2.0	0.3
Manganese	0.2	0.05
Chloride	325.0	250.0
Fluoride	5.2	1.8
Arsenic V & III	0.22	0.05
Total Dissolved Solids	1000.0	500.0

Central treatment of the entire water supply was not considered feasible for many reasons. First, there was a disposal problem with both the arsenic-contaminated sludge from activated alumina column regeneration and the reject brine from the reverse osmosis system. Secondly, the costs of central treatment were considered to be higher than point-of-use treatment. And lastly, central treatment was considered too complicated to be efficiently operated in a community the size of San Ysidro. The results of the study indicated the best solution to be point-of-use treatment with reverse osmosis units.

Since arsenic and fluoride are only considered harmful in water used for human consumption, a point-of-use unit for treatment was needed only for water used for drinking water and cooking. A centrally located large RO unit for only the drinking and cooking water supply for the Village was considered, but there were still concerns about disposal of larger quantities of reject water. There was also doubt that the people would be as willing to use the treated water if they had to travel somewhere to get it. It was decided that the best place to install the treatment units would be in the home's kitchen, preferably under the kitchen sink with a separate faucet on the sink for dispensing the treated water and a small tank under the sink for storing the treated water.

A pilot unit, with a spiral-wound polyamide membrane, was installed in the Community Center to assure the effectiveness of the membrane and the acceptability of the unit to the community. This unit produced about 5 gallons of water a day with a reject rate of about 10 to 20 gallons per day, which was discharged into the user's septic tank. The pilot unit had been in service for about 3 years with little maintenance required.

User Permission

A notification letter was sent to each water customer and a public hearing was held on December 18, 1985 in which the cooperative agreement between the Village and the U.S. EPA was brought before the Villagers to explain the problem with the water quality and to discuss the procedures needed to get the point-of-use devices installed, maintained and tested during the study period. An ordinance was passed by the Village which made the use of Village water contingent upon installation of the RO unit in the water customer's home if the home had indoor plumbing. The ordinance was deemed necessary because point-of-use cannot be a viable alternative to central treatment for a public water system if the water utility operator does not furnish safe drinking water to each water customer. Each water customer also had to sign a permission form to allow the Village to install the unit in their home and to allow access to the unit for testing and maintenance. The permission form was necessary because an ordinance could not give the Village the authority to enter a person's home, only an individual can grant permission to the Village to enter his home.

A few reluctant Villagers did not want the units installed in their homes. The primary reason given was that they did not think they needed them stating, "after all, people in the town had been drinking the water for years and it didn't seem to hurt them". Another reason was the permission they had to give the Village to be able to enter their homes to install, test, and maintain the units. The reluctant few were inevitably "persuaded", however, when they were informed their water was going to be shut off if they did not comply. By the end of January, 1986, 96% of the water customers had signed the forms and by July, 1986, permission was obtained from all 75 of the water customers who were eligible for unit installation. In addition, a few water customers without indoor plumbing and a few non-water customers who were anticipating connection to the Village water system in the future also signed permission forms.

Bid Documents and Evaluation

A Request for Proposal was prepared in which contractors were asked to prepare bid proposals for furnishing approximately 80 units to the Village of San Ysidro, including installation and 14 months of unit maintenance.

The units were required to be under-the-sink models capable of producing a minimum of 5 gallons per day with a storage capacity of 3 gallons. The contractor was required to perform service checks and preventive maintenance on each unit every other month as well as repairs to keep the units operational during an initial 14 month period. Maximum time allowed for service call-outs was three working days. The contractor was required to cover the costs of any household damages resulting from malfunction of the reverse osmosis units during the service period.

Based on the above criteria, each bidder was asked to submit prices for a per unit purchase price, a per unit installation price, and a per unit monthly service charge. Each bidder was also asked to furnish manufacturer's data covering typical installation instructions, construction details, and operating instructions.

Four bid proposals were evaluated. The proposals were evaluated on the factors shown in Table 12 and associated weights as described in the Request for Proposal.

TABLE 12.   FACTORS AND WEIGHTS FOR PROPOSAL EVALUATION
(SAN YSIDRO, NEW MEXICO)

Factor	Weight
1. Construction of Unit	5%
2. History of Similar Installations	5%
3. Proposal Completeness	5%
4. Removal Efficiencies (including amount of water wasted by treatment)	15%
5. Maintenance Required	10%
6. Ease of Maintenance	5%
7. Maintenance Service Contract	15%
8. Price of Units and Installation Cost	20%
9. Maintenance Service Contract Cost	20%
Total	100%

The recommended proposal earned a score of 8.01 on a scale of 0 to 10. The price per unit for purchase, installation, and monthly service was $289.50, $35.50 and $8.60 respectively. This proposal also included the installation of an RO test monitor on each unit. This monitor is an in-line total dissolved solids (TDS) monitor which has a test button the user can push to reveal a green or red light. The monitor can be field set from 50-500 ppm. The green light indicates the TDS in the effluent water is below the preset number. The red indicates that number has been exceeded, and therefore the unit has a potential problem which should be investigated. The total proposal cost based on 80 units and 14 months service was $35,632.00 plus tax.

Installation and Maintenance

The Village Clerk in San Ysidro played a vital role in the coordination of installation and maintenance work for the RO units in the community. The Village Clerk was already the contact for water customers regarding bills and problems, so it was beneficial to continue to use the Clerk to coordinate installations of the RO units and service work with the customers. The Clerk made arrangements with customers to schedule access to their homes for installation and service work and coordinated this with the contractor's service manager.

The contractor was able to reduce the installation time required in each home by preassembling and mounting the filter and module housings to a board. The installer then had only to mount the board, install the faucet, tank, and tapping valve, and connect the tubing to the various parts to achieve an operational unit. The average installation required approximately 30 minutes to complete.

Performance Results

There were up to 72 RO units available on a regular basis for maintenance and testing throughout the study.

A few operational problems were encountered with the units initially. Within the first six months, six modules were replaced because they were not working properly. In addition, approximately 35 initial installations required service within the first six months due to either leaks, the TDS monitor was showing a red light, or there was a problem with water flow.

The initial 14 month maintenance contract was extended to 20 months when the first contract expired. This enabled the Village to have its recently hired maintenance man trained by the contractor's service personnel to perform routine checks and maintenance on the RO units. A breakdown of the total number of service calls performed during the 20 month service period is shown in Table 13. A breakdown of the type, number, and costs of parts replaced is shown in Table 14.

TABLE 13.   BREAKDOWN OF SERVICE CALLS BY TYPE
(SAN YSIDRO, NEW MEXICO)

Type	Number	Comments
Leak	38	8% of leaks reported were not from the RO unit.
Red Light	11	3 of red light calls required a part replacement, others required adjustments only.
Flow Problem	9	2 of reported flow problems were due to low system pressure (25 psi).
Routine Check	217	25% of routine checks resulted in repair or adjustment of unit not identified by customer.
Other	150	Other customer complaints included taste or odor problems, broken faucet handles, noisy air gaps, and reinstallations.
Routine check - No One Home	122	This was 36% of the total routine checks attempted. The actual percentage was probably higher since some "not at home" calls were unrecorded.
Total	412*	

*   Average Number of Calls Per Month = 412/20 = 20.6
Number of Calls Required by Contract = 33-35.   Contract required checks on each installed unit every other month. Number installed varied over contract period.

TABLE 14.   COST BREAKDOWN OF REPLACEMENT PARTS REQUIRED
OVER A TWENTY MONTH PERIOD
(SAN YSIDRO, NEW MEXICO)

Parts Replaced	Number Replaced	Retail Cost/Each($)	Total Cost($)
Module Cap	4	$ 21.10	$ 84.40
Filter Cap	1	15.50	15.50
Drain Clamp	4	6.95	27.80
R.O. Module	16	126.00	2,016.00
Particulate Pre-Filters	59	10.60	625.40
Carbon Filters	14	20.10	281.40
Carbon Post-filters	11	15.60	171.60
Batteries	1	3.00	3.00
Faucet Seal	1	4.56	4.56
Faucet Handle	5	1.96	9.80
Filter Housing	2	27.50	55.00
Module Housing	2	32.50	65.00
Module Seal	1	2.92	2.92
Miscellaneous Fittings	6	1.24	7.44
		Total	$3,369.82

The RO units were very effective in removing arsenic and fluoride from the water. The RO units were effective in removing chloride, iron, manganese, and TDS also, but did not quite meet the removal rates stated in the manufacturer's literature. This was most likely due to the number and concentration of contaminants in the water supply. Figure 15 shows the average monthly arsenic test results from the units and Figure 16 shows the average monthly fluoride test results. Table 15 shows average removal percentage for each of the contaminants during the project period.

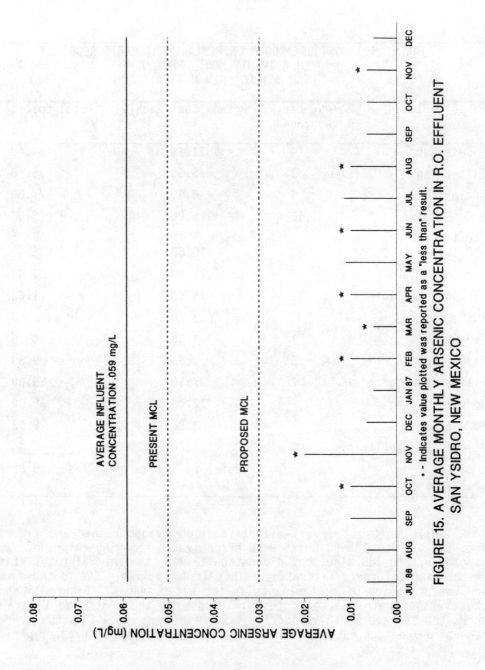

FIGURE 15. AVERAGE MONTHLY ARSENIC CONCENTRATION IN R.O. EFFLUENT
SAN YSIDRO, NEW MEXICO

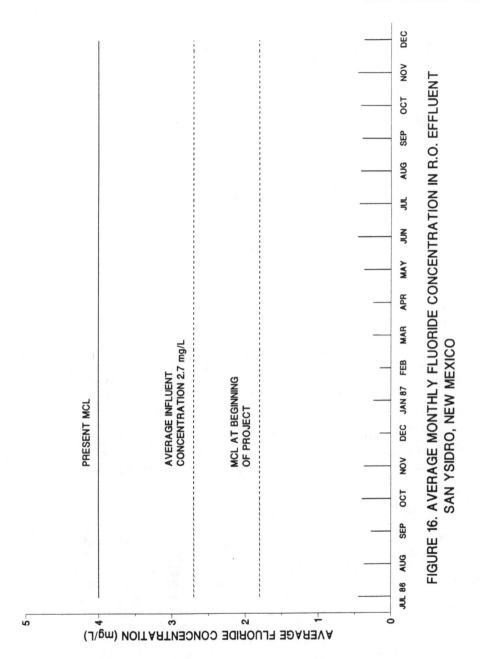

FIGURE 16. AVERAGE MONTHLY FLUORIDE CONCENTRATION IN R.O. EFFLUENT
SAN YSIDRO, NEW MEXICO

TABLE 15.   AVERAGE REMOVAL PERCENTAGE OF TOTAL CONTAMINANTS
(SAN YSIDRO, NEW MEXICO)

Contaminant	Average Influent (mg/L)	Average Effluent (mg/L)	% Removal	% Removal (Manufacturer's Data)
Arsenic	0.059	0.008	86	68/96*
Fluoride	2.7	0.339	87	82
Chloride	91	14.59	84	94
Iron	0.58	0.019	97	--
Manganese	0.09	0.012	87	97
TDS#	780	93	88	94

*   68% removal of arsenic III, 96% removal of arsenic V.
\#   As tested by Contractor's service technicians on routine checks.

Microbiological Concerns

A total of 131 coliform samples were taken from 41 units over a thirteen month period. Five samples were taken from sink taps during the same period. Nine tests from the RO units were positive for coliforms with values ranging from 0 to TNTC (too numerous to count). Ten tests from the RO units yielded non-coliform counts ranging from eleven to TNTC. A total of fifteen of the 131 samples from the units showed evidence of bacterial contamination. Of the five sink tap samples, three were positive for non-coliforms with tests showing counts of 67,137 and TNTC. No test on the distribution system during the project period revealed positive coliform results.

The data collected during this project were inadequate to conclusively evaluate the effectiveness of the point-of-use reverse osmosis unit for removal of bacteria from drinking water. However, presence of non-coliforms before and after the unit initially seemed to indicate that the unit was not significantly removing these types of bacteria. Six units which showed positive coliform test results were reported for servicing. The contractor replaced all three filters, the membrane, and disinfected the tank. Only one unit had a recurrence of a positive coliform test. The second occurrence was five months after the first with two negative tests in-between. Three units were retested after positive test results. The first and second test results are shown in Table 16.

TABLE 16. RETEST RESULTS FOR UNITS SHOWING POSITIVE
COLIFORM TEST RESULTS
(SAN YSIDRO, NEW MEXICO)

Unit Number	First Test	Second Test	Action
14	1 (1)*	75 (TNTC)	Service Unit
44	99 (11)	1 (1)	None
52	5 (1)	1 (TNTC)	None

*   Coliforms (Non-Coliforms) colonies per 100 mL.

The presence of coliforms in the RO effluent water was established when the units were first sampled in December 1987. The first possibility for contamination which was investigated was contamination of the samples during collection. Samples and sample bottle handling and collection techniques were reviewed and emphasized to the Village employee collecting the samples, but sporadic positive coliform samples continued to be seen. The second possibility investigated was system contamination. The Village water supply was chlorinated by a hypochlorination system at the infiltration gallery. The Village had problems in the past with the chlorination system. San Ysidro had been on a boil order for a coliform infraction prior to the installation of the RO units. During the project period no positive coliform tests in the system were obtained, but the monthly system sample was taken from a location very close to the chlorinator. The piping system in the Village was arranged in a three spoke system with the water supply at the hub and the pipes dead-ended at the edges of town. This arrangement could encourage bacterial growth in the stagnant ends of the pipe but the positive test locations did not seem to support this theory. The location of homes with positive results were not in any particular location in the system.

Another possible source of contamination considered was the RO unit itself. Since the polyamide RO membrane utilized in San Ysidro was sensitive to chlorine, the carbon pre-filter was designed to remove any residual chlorine from the water entering the unit. After the carbon pre-filter, the unit was vulnerable to bacterial contamination and growth. Without consistent, adequate chlorination in the water supply system, bacteria could be introduced into the unit. A positive coliform sample was sent to the Cincinnati laboratory for culturing and identification of the coliforms. It was suspected that if the bacteria in the RO units were typical of those found in other studies, they would not be fecal coliforms but coliforms of "non-interest". This identification process revealed, however, that the coliforms were E. coli. This result indicated that there was indeed some contamination or cross-connection of the water system. It was possible that the low system pressure could be inducing backsiphonage from some cross-connections in the individual homes.

Further investigation revealed that the units installed in San Ysidro had not been installed with an air gap on the discharge line from the RO module. The drain line from the RO module was connected directly to the kitchen sink drain. (Every home in San Ysidro had a septic tank system and septic tank problems were common). It was strongly suspected that this was the cross-connection which was causing the positive coliform tests. Discussions with the installer of the units revealed some misinformation regarding the air gaps. The installation contractor felt the air gap was not necessary and that the air gap was just one more place for the unit to develop leaks. In further discussions, however, it was explained to the contractor that without the air gap, especially in San Ysidro where there had frequently been septic system backups, the likelihood of backsiphonage was much greater.

There were no positive coliform tests obtained from the units after the air gap problem was discovered. However, there were only a few months of samples taken after that time.

Conductivity

Testing costs during the project period averaged $1,100 per month. Obviously, neither San Ysidro nor its residents could afford to continue the number or frequency of sampling undertaken during this study. Therefore, an acceptable solution to satisfactory periodic monitoring was investigated with the cooperation of the New Mexico Environmental Improvement Division. Since the units were installed with an in-line TDS meter which actually measures conductivity, the correlation between conductivity and arsenic and fluoride in the water of San Ysidro was studied to see if periodic verification of the conductivity of the RO effluent water would be an acceptable substitute for routine laboratory sampling of the water.

Figure 17 shows a plot of arsenic vs. conductivity on a semi-logarithmic scale and Figure 18 shows fluoride vs. conductivity. From these two figures a conservative rule of thumb for the water in San Ysidro was established. A conductivity reading of 600 micromohs/cm would maintain less than 0.03 mg/L of arsenic and less than 1.0 mg/L of fluoride in the effluent water. Above this number, the unit needed to be serviced, perhaps replacement of the membrane was required. Conductivity is a function of all of the constituents in a water source, therefore, this rule-of-thumb was applicable only to the water source in San Ysidro.

Regulations and Compliance

A number of unique situations developed during the project which revealed the need for modifications of the ordinance at the end of the project period.

The first problem which became evident was the need to deal with commercial and residential users in a different manner. In many cases the small home model unit was insufficient in size to produce and store enough water to meet the needs of a commercial establishment. The commercial

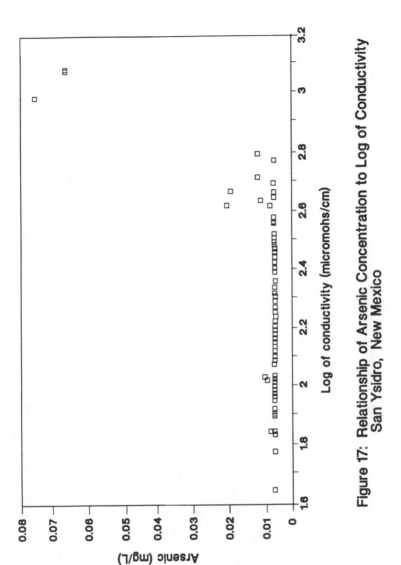

Figure 17: Relationship of Arsenic Concentration to Log of Conductivity
San Ysidro, New Mexico

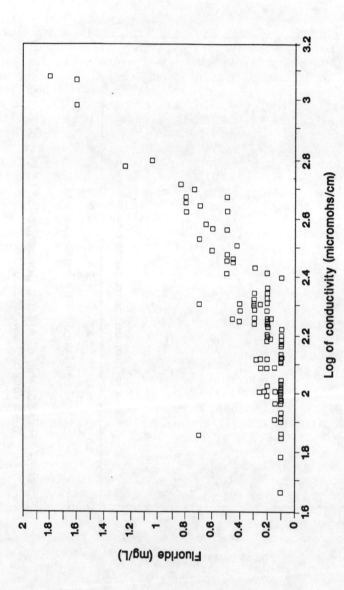

Figure 18: Relationship of Fluoride Concentration to Log of Conductivity
San Ysidro, New Mexico

establishments in San Ysidro were purposefully not considered when developing the project plan because of the complexities involved. The New Mexico Environmental Improvement Division was concerned about the continued resale of water-containing products from these establishments and ordered them to provide safe drinking water which meets regulations to their customers by obtaining a point-of-use treatment system or using bottled water for making coffee, ice, tea, soft drinks, etc. Although the primary responsibility for providing safe drinking water lies with the water utility operator, it was decided to transfer this responsibility to the commercial water user through the new ordinance. This transference served two purposes. First the Village was relieved of the burden of trying to coordinate leasing or purchasing and maintaining various sizes of RO treatment units for the different sizes of businesses in the Village. There was a fluctuation in the number and types of small businesses in San Ysidro at any one time. This fluctuation was evident even during the twenty month project period. Secondly, the new ordinance allowed the commercial water user some latitude in selecting the most economical means for their business to furnish safe drinking water to their customers.

The new ordinance required that the residential units continue to be owned, maintained, and monitored by the Village, but made certain requirements of the individual water customer. The water customer was prohibited from tampering with the unit in any way. In addition, the water customer was required to perform a weekly check of their unit to check for leaks and operate the test switch. Any defect was required to be reported immediately to the Village. If a leak was detected the user was required to turn off the water source to the unit, and the Village was required to repair the leak within two business day after the leak was reported. The user was also required to protect the unit from freezing and prohibited from allowing the unit to drain or stand dry.

Monitoring and sampling of the RO units was to be conducted by the Village. Samples were to be obtained from commercial users as well. If a commercial user did not provide safe drinking water to their customers, the ordinance provided for the installation of an RO unit in the establishment at the commercial user's expense. The Village was required to give notice to water users prior to attempting to check or sample the units. The ordinance provided for penalties ranging from monetary penalties to termination of water service for the user and loss of business permit for a commercial user who did not allow the Village access for monitoring purposes.

Two liability issues were addressed in the new ordinance. First, the Village was held liable for damage to the users home caused by the RO unit with the following limitation. If the damage to the home was caused by a leak that could have been detected in a weekly inspection, and by shutting off the source of water to the unit, the Village would not be liable. Second, the user would be held liable for damages to the unit from tampering, temperature, or allowing the unit to dry out. (The polyamide membrane is destroyed when it is allowed to dry).

An agreement was proposed between the Village of San Ysidro and the State of New Mexico Environmental Improvement Division in which the Village would check the conductivity of each unit with a calibrated hand-held conductivity meter a minimum of once every six months, and would perform a laboratory test for arsenic on each unit once every three years. It should be noted that each unit had the TDS meter set on 200 mg/L which equates to a conductivity of about 500 micromohs/cm. This would provide an additional margin of safety. It was also proposed that the Village substitute one random monthly bacterial sample from an RO unit in place of the monthly distribution system sample.

Costs

Cost for purchase, installation, and maintenance for this project were $289.50, $35.50, and $8.60 per month, respectively. An evaluation of project costs to determine anticipated future costs for the Village of San Ysidro was performed. The purchase price of the RO units was significantly less than the suggested price of $665.00 due to a quantity discount from the manufacturer for the purchase of 80 units.

Based on monthly maintenance estimates, a $7.00 monthly surcharge for the RO unit was recommended. If the cost of the RO unit was included, it would bring the monthly rate to $1,245, still approximately one-half the estimated cost of central treatment. Although the total monthly cost for point-of-use was approximately half of the cost of central treatment, the cost per gallon for treated water should also be compared. The central treatment cost equated to approximately $0.01 per gallons. The point-of-use cost equated to approximately $0.06 per gallon based on a production of 7 gallons of treated water per day.

Conclusions and Recommendations

The reverse osmosis units attained 86 percent arsenic removal, and 87 percent fluoride removal during the 20 month project period. The units were evaluated for removal of chloride, iron, manganese, TDS and bacteria as well.

The reverse osmosis units were effective in removing chloride, iron, manganese and TDS to below the recommended maximum contaminant levels (MCLs), but the removal percentages were approximately ten percent less than those stated in the manufacturer's literature. This discrepancy was more than likely due to the quantity and combination of contaminants in the San Ysidro water supply. The study was inconclusive in its evaluation of removal of bacteria from the water source.

The system of management required for point-of-use treatment is a little more complicated than the "fix it when it breaks" system that most small communities have historically had for their water systems. The devices required a system of scheduled checks and a little more frequent but more simple repairs than most central treatment would require. The recordkeeping for these checks and repairs was estimated to take less than one day per month for a system of 80 units. Maintenance checks and repairs were estimated to require approximately a day and a half per month for a system of the same size.

Maintenance of the point-of-use devices should be done by a water utility employee. The installer of the devices should provide maintenance for the first 2 or 3 months after the devices are installed. During this time, the utility employee should be trained on-the-job by the installer's personnel to perform installation and maintenance repairs. The point-of-use devices are easy to understand, maintain and repair. This should save each water customer at least $1.50 on the monthly charge for the units as well as minimize or eliminate coordination problems between the community and the installer for repairs and maintenance.

The State of New Mexico Environmental Improvement Division cooperated with the Village of San Ysidro to implement the point-of-use treatment system in the community. The Village was required to inspect each unit at least every six months, perform an arsenic test on each unit once every three years, and substitute a random sample from an RO unit each month for bacteriological sampling instead of a system sample. These requirements, in addition to a local ordinance requiring installation of a point-of-use device in each water user's home or place of business, made point-of-use a satisfactory means of providing safe drinking water to the Villagers of San Ysidro.

RADIONUCLIDE REMOVAL CASE STUDIES - SUMMARY

Background

A report by Lowry et al.[12], summarized field data from sites that used GAC POE units to remove Rn from groundwater supplies. One unique aspect of the radon removal process is the fact that the bed is naturally regenerated as radon decays.[13] As a result, the breakthrough exhaustion profile typically seen when GAC is treating other conservative contaminants is not exhibited during radon removal. Information about design, installation, operation, monitoring, performance, gamma exposure rates, and shielding (See Chapter 3) for 121 GAC POE units in 12 states was collected. The locations of the GAC POE units discussed are summarized in Table 17. The majority of units are in Maine and New Hampshire. These units cover a wide range of water quality characteristics.

TABLE 17.   NUMBER OF POE GAC UNITS BY STATE

State	Number	State	Number
Maine	61	Colorado	3
New Hampshire	20	Rhode Island	3
New Jersey	12	Connecticut	6
Kentucky	1	New York	1
Pennsylvania	6	North Carolina	1
Massachusetts	5	Vermont	1

Design, Installation, and Monitoring of GAC POE Units

The GAC POE units were comprised of single pressure vessels housing 1.0 to 3.0 cu ft of carbon, depending upon the model purchased. Model designations were "GAC 10", "GAC 17", and "GAC 30" for bed capacities of 1.0, 1.7, and 3.0 cu ft, respectively. The large majority of installations were done with Model GAC 17. Table 18 is a summary of the relative distribution of unit sizes and their characteristics.

TABLE 18.   RELATIVE USE OF DIFFERENT SIZED GAC UNITS
FOR RADON REMOVAL

GAC Model	GAC cu ft	Vessel Size	Number Installed
Not designated	2.0/3.0	12" X 48"	12/3
GAC 10	1.0	10" X 35"	15
GAC 17	1.7	10" X 54"	72
GAC 30	3.0	14" X 47"	16

The GAC units were all operated in the downflow mode, except for one of the early units put in by a plumber who engaged in water treatment. That unit was installed in the upflow mode by mistake when it was plumbed in backwards. The error was discovered when the unit was included in the recent gamma exposure rate investigation by Lowry[14] and the State of Maine. All GAC units were installed downstream of the existing pressure tank and operated under the normal range of household water pressure. The normal pressure range was 20 to 40 psig or 30 to 50 psig, but some homeowners used pressures up to 60 to 70 psig in rare instances. Normal piping for the installation was 3/4 inch copper. The control valve was a manually operated backwashing/rinse model. The brass body of this valve is widely used and found in many manufacturer's products used in softeners, taste and odor filters, and greensand units. The basket strainer at the bottom of the riser tube had 0.012 inch slotted openings, and it was positioned in a 6 inch deep layer of "pea" gravel. In most instances, the GAC unit was plumbed into the existing line as a side stream with appropriate valves to allow it to be by-passed during service.

Most of the systems did employ an upstream sediment filter and this became a highly recommended option as experience was gained with the operation of GAC units removing Rn. It was used to avoid the accumulation of sediment in the GAC bed, especially for wells drilled in granitic bedrock. Something was needed since regular backwashing, which would normally get rid of the sediment, was detrimental to performance and not recommended. Only 17.8 percent of all installations utilized a water shield or lead shield for gamma exposure rate attenuation. At one site, bricks were used as a shield.

An estimated 60 percent of the installations were made by the homeowner without the assistance of a plumber. The few problems that arose during the installation of the units were usually associated with a plumber or water treatment dealer that had failed to read the instructions, or had ignored them altogether. The problems were usually related to conducting the initial filling and backwashing of fines at a higher than recommended rate, with subsequent packing of the control valve internals with GAC. This either caused an incomplete seal (a partial by-pass of untreated water) or damage to the valve o-rings when they forced closed against the GAC. In a few cases, the by-pass valves were left open and the unit only treated a small fraction of the water until the problem was identified and corrected by properly positioning the valves. A final problem that occurred when units were shipped to northern New Jersey was one related to the actual shipping. One specific trucking terminal removed secured upright GAC units from their original pallet and transported them in a horizontal position. It was first suspected that this might have been the reason that some of those particular units had not performed as well as the norm, since the horizontal shipping may have disturbed the layer of gravel underdrain in the unit. However, it was determined later that the performance problem was probably related to the local water quality, rather than to the unit or the GAC.

Once the GAC unit was properly installed and commissioned, it was essentially maintenance free. The sediment filter, when used, was typically in the 30 to 50 micron range and required replacement or washing approximately twice per year. Occasionally, a water supply was encountered that had abundant sediment and the filter would need cleaning or replacement as often as once per month. One extraordinary water supply was encountered that required sediment filter cleaning once per week.

Backwashing of GAC units was recommended only if the hydraulic capacity of the unit became noticeably diminished, as indicated by a significant drop in water pressure at the tap. Regular backwashing once per week had been reported to cause a lower overall removal efficiency and was not believed to be needed if a sediment filter was in place.[13] This was seen by experience in the field, as very few cases have been reported where loss of hydraulic capacity has been an issue. However, a few units have appeared to experience trouble over the years when a sediment filter was not used. An example was one of the units included in the gamma exposure rate study mentioned previously.[14] This unit had never been backwashed during its life of 43 months and a loss in pressure was noted. Unfortunately, this was the unit that had been operated/installed in the upflow mode and it could not be effectively backwashed. In general, a backwashing recommendation of once per year has been adopted for those homeowners that have sediment filters and feel a need to backwash. In summary, backwashing has not been an issue when a sediment filter has been used.

Performance Results

In actual field operation, there were units that achieved equal to or greater than predicted performance; however, field units taken altogether give an overall removal of something less than theoretical. This is clearly indicated by the average numbers for the actual removal shown in Table 19, and by the histogram of unit performance shown in Figures 19 and 20.

TABLE 19.   EXPECTED VS. ACTUAL PERFORMANCE FOR POE GAC UNITS
FOR RADON REMOVAL

Performance GAC Model	Flow gpd	Empty Bed Contact Time hr.	Expected %	Actual %
GAC 10	157	1.14	96.7	90.7
GAC 17	157	1.94	99.7	92.5
GAC 30	157	3.43	99.99+	98.6

Note: Number of units in "actual" categories are 12, 59, and 12 for No. 10, 17, and 30, respectively.

- Approximately 84 percent of the 121 POE GAC systems in the field were achieving removals of greater than 95 percent. Approximately 94 percent of all units achieved greater than 90 percent Rn reduction.

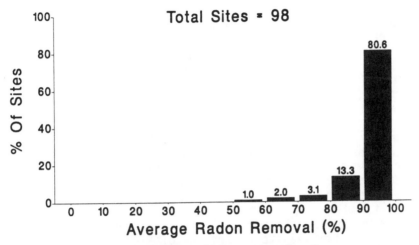

Figure 19:   Histogram of Average Performance
             for all POE GAC Units

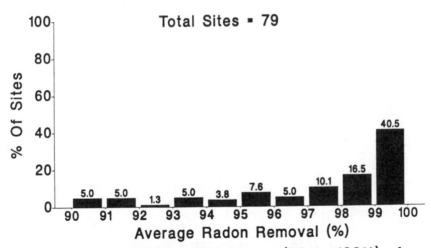

Figure 20:   Expanded Histogram (90 to 100%) of
             Average Performance for POE GAC Units

- Approximately 6 percent of the POE GAC systems have experienced a premature failure that was believed to be water quality related. The problem of premature failure was clearly associated more frequently with particular regions within specific states.

The numbers for actual removal in Table 19 are simply the average removal data, taken in its entirety, without regard to any specific knowledge that may exist for the units or the data. Possibilities for errors in sampling, possible partial by-pass due to equipment problems, or improper plumbing, etc. were not investigated in every case where removal was less than expected. Higher than estimated water use could also have played a role. While these elements to the overall data set probably exist, it is believed that their impact on the performance numbers are very small. Rather, unknown water quality factors at the specific sites are believed to be responsible for the lower performance at some of the sites.

Service Life

At the present time there are not enough data to predict the long term life of the Rn adsorption/decay steady state. Each GAC unit must be considered on a case-by-case basis. A few systems prematurely fail and cannot be said to be entirely effective for even several months. Others have continued to show theoretical removal efficiencies for extended periods without any signs of deterioration of the adsorption/decay steady state. Other factors such as Pb-210 and its progeny buildup in the bed may dictate the service life of a given GAC bed in a given state.[15]

As a final analytical tool for the data collected to date, units that had sufficient data over three or more years were analyzed for their long term performance. The data for three of the sites are summarized in Figures 21-23. There are no general trends in these data and it is clear that continued steady state removal has been present over the years. Figure 22, shows the most variation but it must be noted that even though a significant monitoring effort is represented, the data are still limited. It is not known for Figure 23 if a downward trend is a reality or just reflective of variable flow or influent water quality.

Costs

The costs associated with GAC installations during 1986 and 1987 are summarized in Table 20. Costs include the GAC unit, sediment filter and gamma shield, if present, and the installation parts and labor. Operation costs have not been estimated as the ultimate GAC bed life for these field units is not known and possible regulatory controls for limiting the GAC bed life based upon a specific accumulation of Pb-210 are non-existent at the present time. Should limitations for Pb-210 be imposed, the bed life might be as short as a few months, making the operational costs prohibitively high.[15]

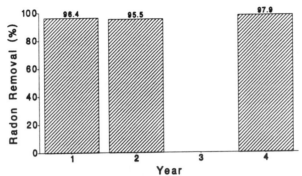

Figure 21: Long Term GAC Performance for Radon Removal, North Sutton, NH

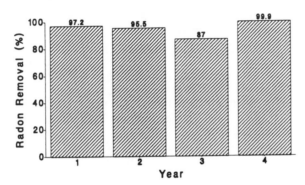

Figure 22: Long Term GAC Performance for Radon Removal, Deerfield, NH

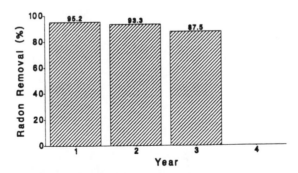

Figure 23: Long Term GAC Performance for Radon Removal, Gorham, NH

TABLE 20.   COSTS ASSOCIATED WITH POE GAC INSTALLATIONS
FOR RADON REMOVAL (1989 DOLLARS)

GAC Model	GAC Unit	Sediment Filter	Water Shield	Installation	Total
GAC 10	$600	$50	$25	$100	$775
GAC 17	750	50	90	100	990
GAC 30	450	50	125	100	1,225

In general, the O&M costs for the POE GAC installations have been negligible. For the few units that have failed in the field, bed replacement appears to be needed within a 6-month period. The bed replacement is estimated to cost approximately $130 per cu ft of GAC, including labor. The large majority of systems have never had a bed change. In fact, none of the units have had a routine bed replacement. One unit that had failed in New Jersey was replaced on two separate occasions in an unsuccessful effort to determine the cause of failure. Another unit was upgraded by changing the bed to a superior GAC product. On a long term basis, Rn analyses to determine the performance of a GAC unit are recommended semi-annually and would cost about $40/year in 1989 dollars.

Conclusions

While POE GAC installations were found to be effective, the use of GAC for Rn removal may be limited in the future to wells containing less than 5,000 to 10,000 pCi/L. This would result if the private residence desired to achieve the new MCL for Rn, which is expected to be set between 200 and 2,000 pCi/L. In these applications, POE GAC units will be considerably more economical than aeration systems, and have minimal gamma exposure rate fields. It is estimated that hundreds of these units are in service today as a result of their use for the removal of other water contaminants.

While POE GAC units have limitations where the supply has relatively high Rn, it is estimated that in excess of 90 percent of all wells fall below 5,000 pCi/L. Therefore, if widespread general application of treatment for wells above the new stringent MCL occurs, GAC systems may play the major role in this treatment application.

Additional Radionuclide Removal Studies

Research undertaken between the USEPA and University of New Hampshire Environmental Research Group evaluated radon removal in small community water supplies using full-scale GAC, diffused bubble aeration, and packed tower aeration.[16] Even though full-scale systems were evaluated, several of the conclusions and recommendations summarized below are still pertinent to these technologies used in the home.

Granular Activated Carbon

1. When designing a GAC system to remove radon, it is imperative to have good data on water flowrates and influent radon activities at the site. These data are major inputs into the design model for determining the volume of GAC required. As observed in this study, variations in flowrate and influent activity may be substantial and will, if underestimated, lead to inadequate system design and effluent radon activities which exceed the design goal.

2. The steady state adsorption-decay constant (K_{ss}), which is a critical component of the GAC design model varied over time, was site specific, and differed by 22% to 89% from the value previously reported in the literature. The factors which will most likely influence K_{ss} for a given site are flowrate and radon concentration which effects mass transfer and adsorption kinetics, and raw water quality which effects adsorption capacity. Therefore, pilot scale testing should be performed at each site to determine the appropriate K_{ss} to be used in the design model.

3. The data indicated that the GAC units retained uranium 238 and 235, and radium-226 in quantities (pCi/kg) high enough to classify the carbon as a low level radioactive waste according to State of New Hampshire regulations. The units also retained lead-210 which is not currently regulated in the State of New Hampshire. Therefore, the GAC disposal costs could increase the production cost of a GAC system. Concentration of uranium, radium, and lead by GAC systems appears to be related to raw water quality (e.g., pH and alkalinity).

4. The GAC units accumulated iron, manganese and particulates (turbidity). In addition, there were significant numbers of microorganisms growing on the carbon. As a result, GAC systems may require periodic backwashing to prevent significant headloss development. In certain water supplies, concentrations of iron, manganese and particulates may be high enough to warrant frequent backwashing which would increase operating costs and reduce the hydraulic capacity of the system. In addition, previous studies have shown that frequent backwashing may decrease radon removal efficiency. Therefore, in these cases, pretreatment for iron, manganese and particulates is recommended to decrease the backwashing frequency.

5.   Gamma/beta emissions measured at the surface of the GAC units were substantially greater than background measurements. Some form of shielding will be required to lower these emissions to acceptable levels.

6.   Further studies should be conducted to examine the effects of backwashing and raw water quality parameters (e.g., organics and iron) on radon removal efficiency in GAC systems. In addition, the factors affecting the retention/release of radionuclides such as uranium, radium and the radon progeny should be investigated.

Diffused Bubble and Packed Tower Aeration

1.   When designing an aeration system to remove radon, it is imperative to have good data on water flowrates and influent radon activities. These data are major inputs into the design model for determining hydraulic detention time, air flowrate and diffuser configuration. As observed in this study, for small community supplies, variations in water flowrate and influent activity may be substantial and will, if underestimated, lead to inadequate system design and effluent radon activities which exceed the design goal.

2.   The percent radon removal in a diffused bubble system is directly proportional to the global liquid mass transfer coefficient (K_La), the detention time and the concentration gradient ($C_{liquid} - C_{gas}$). These findings are analogous to those reported for other more conventional contaminants. Since K_La, detention time and concentration gradient are system specific, extrapolations of performance data obtained at one site should not be made to systems with other configurations/diffusers, packing material, low influent activities and/or those required to meet a more stringent MCL. Therefore, pilot-scale studies may need to be conducted at each site to obtain the required design information. High influent Rn levels may require more packing height than is possible in an individual home and preclude use of this design.

3.   Off-gas radon activities at the Derry, NH site ranged from 3,361 to 18,356 pCi/L in the air exiting the diffused bubble system and were 10^4 to 10^5 times higher than ambient air levels. The off-gas radon activities ranged from 2,410 to 21,200 pCi/L in the air exiting the packed tower and were 10^4 to 10^5 times higher than ambient air levels. Therefore, the impact of the off-gas on ambient air quality should be considered.

4.   Though precipitation of iron and manganese was not observed during the short-term diffused bubble testing, it is well documented that this can occur in aeration systems treating

groundwater, resulting in operational problems. Therefore, raw water quality should be monitored to determine whether pretreatment is required.

5.   In northern climates, aeration systems, including the blower intake, are typically located inside the pumphouse. Since it has been shown in this study that the radon activity in the pumphouse air may be high ($\approx$100 pCi/L at Mount Vernon, NH) use of this air may significantly affect the ability of the aeration system to meet a stringent MCL due to mass transfer limitations. Therefore, the radon activity of the influent air should be considered in the design and an outside air intake may be required.

VOC REMOVAL CASE STUDIES

In the Village of Silverdale, PA, 49 POU activated carbon (GAC) devices, representing products from several manufacturers, were installed and monitored for 14 months of operation for control of volatile organic chemicals (VOCs), most notably trichlorethylene (TCE) and tetrachloroethylene (PCE). In the Lake Telemark subdivision of Rockaway Township, NJ, the township health department and a manufacturer of POU GAC devices began a pilot demonstration in 1981 by installing and monitoring devices in 12 homes with wells contaminated with organics. Performance verification and review of cost data were included in this study[17]

The major VOCs in Silverdale's water supply were TCE and PCE. POU devices reduced concentrations of these contaminants to nondetectable levels (<1 μg/L) in 87 percent of the samples collected over 14 months. The mean volume treated during this period was 340 gal; maximum volume treated was 1,130 gal. Devices were still in operation at the end of the study.

Breakthrough, defined as detection of the same VOC in consecutive postdevice samples from the same site at concentrations above 1 μg/L, did not occur for any device for TCE or PCE during 14 months of sampling. However, trace concentrations of VOCs were detected intermittently in postdevice samples from each model type; concentrations were generally below 5 μg/L. The most frequently measured postdevice VOC was chloroform. Although the mean influent chloroform concentration was 12 times less than the mean TCE concentration, chloroform may break through before TCE. This is supported by isotherm data typical for activated carbon.

The capital cost for POU activated carbon treatment in Silverdale ($298 in 1985 dollars) was an average cost of purchasing devices from several manufacturers (in quantity) and equipping them with product water meters. Maintenance costs included an average monthly repair cost per site of $1.43 in 1985 dollars. Some POU devices required no maintenance during the study.

In the Lake Telemark subdivision of Rockaway Township, 12 POU activated carbon devices were installed on private well water supplies in October 1981. Only one of 21 postdevice samples collected from October 1982 through October 1983 contained detectable VOCs (4 μg/L TCE and 2 μg/L PCE). Eight sites were sampled during the 24 months of operation with no detectable VOCs in effluent samples. After 2 years of service, the average cumulative volume treated was approximately 1,650 gal, based on readings taken from a flow indicator on the device. A device sampled after reaching its estimated treatment capacity of 2,000 gal produced water with no detectable VOCs.

Equipment costs of POU activated carbon devices in Rockaway ($25 in 1985 dollars) were negotiated by the community during an initial phase of the pilot demonstration. No maintenance was reported during the 2 year pilot demonstration period. A summary of results from demonstrations of POU activated carbon devices in Silverdale and Rockaway Township appears in Table 21.

Microbiological Concerns

Bacteriological Sample Results

Standard Plate Counts (SPCs) from the GAC sites indicated microbial colonization of the carbon bed. In Silverdale, unflushed postdevice samples had mean densities two orders of magnitude higher than corresponding predevice samples. If one liter of water was flushed from the line before sampling, postdevice samples had mean densities only one order of magnitude higher than predevice samples. Samples of water collected after 2 min of flushing had SPC densities comparable to samples of water from the distribution system. In Rockaway, flushing and disinfecting the tap reduced SPCs by one order of magnitude. Data collected during the study did not indicate colonization of activated carbon devices by coliform organisms. Positive coliform results in Silverdale were obtained from 4 of 176 postdevice samples collected from flushed, disinfected taps. Postdevice resamples were negative for coliform organisms. No coliforms were detected in postdevice samples collected in Rockaway.

TABLE 21. POU ACTIVATED CARBON STUDIES

Participating Sites	Service Area Type	Mean Treated Water Use (gpd)	Trichloroethylene		1,1,1-Trichloroethane		1985 Dollars	
			Predevice (mean μg/L)[1]	Postdevice (mean μg/L)[1]	Predevice (mean μg/L)[1]	Postdevice (mean μg/L)[1]	Capital Cost[2]	Customer Cost ($/month)[3]
Silverdale, PA	Central System with single family homes	1.0	80	<1	1	<1	289	5.98
Rockaway Township, NJ	Private wells at single family homes	2.3 est	125	<1	92	<1	255	4.23

1 Samples containing <0.001 mg/L were assigned a value of 0.0009 mg/L for calculation of the mean.

2 Average of five manufacturers; includes equipment + installation costs.

3 Capital, amortized at 10% for 20 years + maintenance.

ARSENIC REMOVAL CASE STUDIES - SUMMARY

Reverse osmosis (RO) ion exchange (IEX) and activated alumina (AA) treatment technologies for arsenic removal have been field tested in Fairbanks, Alaska and Eugene, Oregon in addition to San Ysidro, New Mexico.[18]

The RO units installed at each location were low-pressure home RO systems. The units as purchased were designed to produce about 3-5 gal (11-19L) of drinking water per day. These units were also designed to operate with source water pressures ranging between 20 and 100 psi (138 and 160 kPa) with a product-to-reject-water ratio of about 1 to 10. Water was fed to the RO systems from a tap on the influent line to the household plumbing system. The average influent pressure was 40 psi (276 kPa). The water entering the RO system was prefiltered through a 5-μm cartridge filter before it passed through a cellulose-acetate, spiral-wound membrane. Permeate was collected in a storage tank and then passed through a polishing carbon cartridge filter prior to the effluent valves.

After two years of operation, a second type of RO system was installed at one location. The new RO system was similar to the old unit, except that a booster pump was added to increase the operation pressure. The system operated at 195 psi (1,346 kPa) with a product-to-reject-water ratio of 1 to 3.

The IEX tanks used in this study were cylindrical and measured 46 in (117 cm) high by 9 in (22 cm) in diameter. The tanks were filled with 1 cu ft (0.028 m^3) of a strong base anion exchange resin. Flow rate through the cylinders was controlled by the effluent valving and initially set at 1 gpm (3.78 L/min). At this flow rate, the surface loading rate of the tank was 2.7 gpm/sq ft (0.95 L/min/m^2) and the empty bed contact time was 7.5 min. The actual contact time was probably greater because the effluent valves were only opened for 1 min by the timers, and the water was undisturbed in the tanks until the next valve-opening period. The contact time was occasionally shorter when sampling valves were opened manually and the effluent was allowed to flow to waste for a short period of time.

The AA tanks were identical to the IEX tanks with the exception of the media. The AA medium was granular activated alumina (1 cu ft [0.028 m^3]). The medium was pretreated (in the tank) by passing a sodium hydroxide solution through the tank, rinsing with clean water, and then treating with dilute sulfuric acid to lower the pH of the AA. The operation and water throughput of the AA tanks was identical to those described for the IEX tanks.

All four homes used groundwater (private wells) for their drinking water. The source-water quality at the four sites varied. The arsenic concentration was the most variable parameter and ranged from <0.005 mg/L to >1.1 mg/L during this study. The typical source-water concentrations for various parameters are listed in Table 22.

TABLE 22.   SOURCE WATER QUALITY FOR
ACTIVATED ALUMINA STUDY

Parameter	Oregon Houses (mg/L)		Alaska Houses (mg/L)	
Calcium	18	19	22	8.9
Magnesium	5.3	5.5	10.6	9.3
Alkalinity	151	206	108	56
Turbidity (NTU)	0.43	0.24	0.48	0.32
Chloride	<10	<10	<10	<10
Sulfate	<15	<15	<15	<15
pH	8.3	8.3	8.0	7.4
Iron	0.24	0.18	<0.1	0.20
Sodium	40	62	6.0	4.4
Arsenic	<0.005-0.28	0.005-0.32	0.25-1.08	0.22-1.16

Two of the water quality parameters listed (iron and sulfate) were low. Both of these compounds may cause problems in the removal of arsenic, and their concentrations must be determined when removing arsenic. Iron compounds will clog and foul IEX resins and AA, thereby lowering the removal capabilities of each medium or possibly clogging the units and reducing the water throughput. Sulfate is preferred over arsenic by an anion resin and also interferes with removals by AA. The low sulfate concentrations provided good removal conditions by both processes.

All three treatment techniques (IEX, AA, and RO) were shown to be capable of removing arsenic from water. The low-pressure RO systems worked well in lowering the arsenic concentration and would be successful in treating water supplies in which the arsenic concentration did not exceed 0.10 mg/L (this value is based solely on the observed 50 percent arsenic rejection). The high-pressure system worked very well in removing arsenic but required the addition of a booster pump to achieve high pressure. The small amount of water produced by the RO systems would limit their use to treating water for drinking and cooking. Another disadvantage of the RO systems was the waste stream, i.e., the production of 9 gal (34 L) of reject water for every gallon of treated water. One advantage of RO is that the concentration of many other components are lowered while the arsenic is being removed; thus a better overall quality of water is produced. Also, because arsenic is not accumulated in an RO system, a spent membrane would be easily disposed.

The IEX units worked well when properly pretreated and successfully treated water containing as much as 1.16 mg arsenic/L. The two 1-cu-ft units in Alaska treated more than 11,000 gal (42 m^3) and 16,000 gal (60 m^3), respectively, and were disconnected at termination of the project before media exhaustion. The Oregon units also worked well when the IEX resin was pretreated properly for anion exchange.

The AA tanks also worked well on the Alaskan waters (after problems were corrected), and a conservative media capacity of 1 mg arsenic/AA could be expected. Thus, if the raw water arsenic concentration was 0.3 mg/L, an AA tank containing 1 cu ft of media should be able to treat 22,000 gal (83 m^3). The poorer removals seen in Oregon may have been due to poor pretreatment. Iron and sulfate in the water would also lower the arsenic removal capacity of the alumina and should be monitored. Table 23 provides a summary of volume of water treated, arsenic removal and days of operation.

Although all three treatment techniques will work in removing arsenic, care must be taken in preparing and installing the treatment systems. The media must be pretreated correctly to ensure arsenic removal, and monitoring must be done periodically after installation to confirm that arsenic is actually being lowered to below the MCL.

TABLE 23.   DATA SUMMARY FOR ARSENIC REMOVAL

Site	Treatment Unit	Water Treated Until MCL Exceeded (gal)	Days on Line Until MCL Exceeded	Amount of Arsenic on Media (mg)
Oregon				
Site 1				
0.026-0.22 mg/L,	RO-1	322	225	NA*
Arsenic				
	RO-2	NA	100	NA
	IEX-1	5,230	268	232
	IEX-2	6,690	314+	999
	AA-1	7,768	406	2,470
	AA-2	7,130	371+	2,610
	AA-3	6,342	314+	1,290
Site 2				
0.34-1.08 mg/L,	RO-1	NA	180	NA
Arsenic	RO-2	NA	14	NA
	IEX	16,254	1,471+	23,100
	AA-1	0	0	606
	AA-2	15,427	1,226+	25,200
Alaska				
Site 1				
0.22-1.16 mg/L,	RO-1	NA	90	NA
Arsenic	RO-2	NA	35	NA
	IEX	11,858	680+	21,600
	AA	10,784	680+	20,700
Site 2				
0.05-0.35 mg/L,	RO-1	NA	119	NA
Arsenic	RO-2	NA	650+	NA
	IEX-1	NA	0	1,520
	IEX-2	20,935	427	6,988
	IEX-3	9,410	350+	4,897
	AA-1	NA	0	2,850
	AA-2	18,557	427+	6,369
	AA-3	7,538	350+	4,845

* NA-Not available
+ Taken off line before MCL exceeded

URANIUM REMOVAL CASE STUDIES

A uranium removal study was conducted by USEPA in Colorado and New Mexico using 12 POU IEX tanks containing 0.25 cu ft (7.0 cm^3) of a strong-base anion.[19]

Six of the units were set to operate intermittently to simulate point-of-use-type units, while the other six were operated in a con-tinuous-flow setting. All 12 had controls to regulate the flow to 0.25 gpm (0.016 L/s). Because of a water shortage, one of the continuous-flow units had to be discontinued. All units were routinely sampled to deter-mine total uranium removal and were metered to measure the volume of water treated (Table 23). The influent uranium concentration to these units was 22-104 μg total U/L. In all but three of the units, the effluent water contained <1 μg U/L when the systems were shut off after approximately two years operation. Effluent water samples from units C, E, and G displayed uranium concentrations exceeding 1 μg/L at 10,000, 13,000, and 8,000 bed volumes, respectively. After 12,000 bed volumes of water were treated with unit G, the effluent uranium concentration exceeded the influent level. The IEX tanks cost $125 each. Installation and O & M costs were not available. As indicated in Table 24 the anion exchange systems performed well in lowering the uranium concentrations in a number of different water supplies.

TABLE 24.   POINT-OF-USE URANIUM REMOVAL

Unit	Raw Water Total Uranium Concentration μg/L	Quantity of Water Treated at Shutdown Bed volumes	gal (L)	Uranium Removal at Shutdown percent
A (I)*	22	9,485	17,740(67,146)	99.5
B (I)	30	25,182	47,098(178,266)	100
C (C)	23	24,820	46,420(175,700)	0+
D (I)	52	2,812	5,260(19,909)	100
E (C)	52	34,690	64,879(245,567)	13.5+
F (C)	52	21,280	39,800(150,643)	100
G (C)	35	11,929	22,310(84,443)	0+
H (C)	28	63,200	118,200(447,387)	98.2
I (I)	100	13,970	26,128(98,894)	99
J (I)	40	20,152	37,690(142,657)	100
K (I)	104	7,897	14,770(55,904)	100

*(I) - Intermittent-flow unit; (C) - Continuous-flow unit
+ Medium saturated at termination of study

FLUORIDE REMOVAL CASE STUDIES - SUMMARY

Several Illinois and Arizona communities installed POU devices for fluoride reduction.[20] Arizona communities using POU activated alumina (AA) devices for fluoride reduction included Thunderbird Farms and Papago Butte Ranches, where separate distribution systems are provided for irrigation and domestic water. A portion of the domestic water was bypassed and treated with AA for potable uses. Domestic water boards for both communities provided installation, monitoring, and maintenance of treatment devices. POU treatment with AA for arsenic and/or fluoride reduction was studied at two Arizona institutions, the Ruth Fisher Elementary School located near Tonopah and the You and I Trailer Park located new Wintersburg.

At the three Illinois project sites, the public water systems were supplied by well water with high fluoride, alkalinity, and dissolved solids. Project demonstrations included installation and monitoring of 10 POU AA devices in Parkersburg and 40 POU AA devices in Bureau Junction. In Emington, 47 low-pressure POU reverse osmosis (RO) devices were installed and monitored for eight months. These sites were the first applications of POU fluoride reduction at the community level in Illinois. Table 25 provides a summary of the field sites.

TABLE 25. LOCATIONS WHERE FLUORIDE REMOVAL WAS EVALUATED

Site	Treatment Approach	Treatment Process	Application
Thunderbird Farms, AZ	POU	AA	Fluoride Reduction
Papago Butte, AZ	POU	AA	Fluoride Reduction
Ruth Fisher School, AZ	POU	AA	Fluoride Reduction
You & I Trailer Park, AZ	POU	AA	Fluoride & Arsenic Reduction
Parkersburg, IL	POU	AA	Fluoride & Arsenic
Bureau Junction, IL	POU	AA	Fluoride & Arsenic
Emington, IL	POU	RO	Fluoride & Dissolved Solids Reduction

At Thunderbird Farms, several POU AA devices reduced influent fluoride levels effectively for periods exceeding 2 years. Other AA devices operating at Thunderbird Farms had shorter service lives, attributed to media cementing and/or short-circuiting. At the You and I Trailer Park, raw water fluoride concentration was 15.7 mg/L and arsenic was 0.086 mg/L. A POU AA device effectively treated 2,500 gal (330 bed volumes) before fluoride breakthrough, demonstrating the highest exchange capacity observed for POU devices (2,300 grains/ft^3). Influent arsenic and silica concentrations at Arizona POU sites generally were reduced to nondetectable levels beyond fluoride breakthrough.

The effect of raw water alkalinity is demonstrated in the data from Illinois sites, which included a pilot demonstration in Emington. The higher alkalinity at Parkersburg caused fluoride breakthrough at Bureau Junction, with the lowest alkalinity, was not observed until 350 bed volumes. Part of the reduced capacity at Emington was attributed to the accelerated flow (370 gpd) during the pilot study.

Maintenance costs for POU AA devices were based on replacing the alumina cartridge when treated water fluoride levels reached the Maximum Contaminant Level (MCL). For Arizona, the MCL was 1.4 mg/L, and for the Illinois communities, it was 1.8-2.0 mg/L. Summaries of results from Arizona and Illinois POU AA sites appear in Table 26 and Table 27, respectively.

POU RO systems installed in Emington used a spiral-wound polyamide RO membrane operated at line pressure. Pretreatment included granular activated carbon (GAC) followed by a 5-micron prefilter. Product water was stored in a 2-gal pressurized tank. Reject water was bled through a capillary tube to the home drain line. Product water from the storage tank passed through a GAC polisher before being dispensed.

Fluoride rejection averaged 86 percent, with total dissolved solids rejection (TDS) averaging 79 percent. Relatively large ranges of rejection percentages were observed for all analytes. This phenomenon did not correlate with site, use rate, or collection date, but appeared to be due in part to a pressure drop across the prefilter assembly. Flow rates were measured for several RO devices during a site visit. Ranges of product and reject flow rates were 1.3 to 4.4 gpd and 16.1 to 27.8 gpd, respectively. Water temperatures and pressures (measured at hose connections) did not correlate with flow rates. Iron deposits in the well and distribution system fouled some GAC prefilters, creating head loss across the pretreatment assembly. One GAC prefilter which had been fouled with iron deposits was removed, flushed, and reinstalled. The resulting 33 percent production rate and 29 percent increase in solids rejection, implied a constant flux of solids across the RO membranes, i.e., more water was produced for essentially the same mass of solids, resulting in higher quality water.

The capital cost for POU RO at Emington ($540 in 1985 dollars) was an average of several manufacturers' quotes for devices, with and without pressurizing pumps, based on purchase of 40 to 50 units. The average installation cost per unit of $68 (performed by an equipment dealer) was included.

TABLE 26. ARIZONA POU ACTIVATED ALUMINA STUDIES

	Thunderbird Farms	Papago Butte	Ruth Fisher School	You & I Trailer Park
Participating Sites	8	1	1	1
Service Area Type	central system with singe family homes	subsystem for several families	institution	institution
Influent Fluoride (mg/L)	2.6	2.6	4.4	15.7
Influent Alkalinity (mg/L as $CaCO_3$)	200	200	80	40
Mean Treated Water Use (gpd)	1.4	18.5	8.5	5.5
Volume to Breakthrough[1]				
(gallons)	>1,540	9,500	1,000	2,500
(bed volumes)	>410	1,270	270	330
Costs				
Capital ($)	225	350	360	230
To Customer ($/month)[2]	4.44	4.60	12.00	6.27

1   Defined as the point where postdevice fluoride concentration reached the local MCL.
2   Capital, amortized 10% for 20 years + maintenance.

TABLE 27.   ILLINOIS POU ACTIVATED ALUMINA STUDIES

	Parkersburg	Bureau Junction	Emington
Participating Sites	10	40	1
Service Area Type	central system with singe family homes	central system with single family homes	pilot study
Influent Fluoride (mg/L)	6.6	6.0	4.5
Influent Alkalinity (mg/L as $CaCO_3$)	1,000	540	880
Mean Treated Water Use(gpd)	0.6	0.8	370
Volume to Breakthrough[1]			
(gallons)	400	1,300	700
(bed volumes)	110	350	190
Costs			
Capital ($)	273	285	273 est.
To Customer ($/month)[2]	6.23	4.25	5.38 est.

[1]   Defined as the point where postdevice fluoride concentration reached the local MCL.
[2]   Capital, amortized 10% for 20 years + maintenance.

Costs for central RO treatment at Emington were estimated by soliciting a quote; they included approximately $60,000 in 1985 dollars for a central RO system (including mechanical and electrical installation) and $60,000 in 1985 dollars for a concrete block building. Estimated operating costs per 1,000 gal of product water included chemicals ($0.10), power for pumps ($0.36), membrane replacement every 5 years ($0.18), and prefilter cartridge replacement ($0.02). All costs were in 1985 dollars. Monthly customer costs were based on the design flow of 16,500 gpd. A summary of the Emington RO demonstration appears in Table 28.

Microbiological Concerns

Bacteriological samples collected at AA POU sites indicated microbial colonization of the alumina bed, though not as great as with activated carbon. At the Arizona sites, slight increases in SPCs through AA devices were observed. Flushing reduced SPCs by a small margin. No coliforms were detected in Arizona AA postdevice samples. In Bureau Junction, postdevice SPCs were highest when devices were first placed in operation, and decreased with use. There was no evidence of colonization of AA devices by coliform organisms. Out of 153 samples, coliforms were detected in 9 predevice samples and 4 postdevice samples. One unit maintained consistent positive coliform results and was removed from service. Resamples from other units were negative for coliforms. In Parkersburg, postdevice SPCs were highest when no influent chlorine residual was detected. Flushing and disinfecting taps reduced postdevice SPCs by an order of magnitude. No coliforms were detected in 80 Parkersburg postdevice samples.

SPC results from Emington (RO/GAC) demonstrated an order of magnitude increase through the treatment system. Limited sampling from stages in the RO/GAC system indicated that most bacterial growth was occurring in the GAC polisher. Of 92 samples, coliforms were detected in 4 predevice and 11 postdevice samples. One site was resampled twice before postdevice samples were clear, and another site required disinfection of the RO system twice before resamples were acceptable. Resamples from other units were acceptable.

Emington, Illinois Revisited

The RO units were, in general, operating satisfactorily. Only a couple were removed by the homeowners believing them to be unnecessary. The main drawback seemed to be the low water output producing the 3 gallons per day as rated. Several homeowners opted for bottled water to supplement their needs, purchasing up to 20-30 gallons per month at a cost of one dollar per gallon.

TABLE 28.   EMINGTON, ILLINOIS, POU REVERSE OSMOSIS STUDY

Participating Sites	47
Service Area Type	central system with single family homes
Mean Treated Water Use (gpd)	0.8
Mean Flow Rates (gpd)	
Product Water	2.9
Reject Water	22.5
Fluoride (mean mg/L)	
Predevice	4.5
Postdevice	0.6
Total Dissolved Solids (mean mg/L)	
Predevice	2,530
Postdevice	520
POU Treatment Costs	
Capital ($)[1]	540
To Customer ($/month)[2]	12.48
Estimated Central Treatment Costs	
Capital ($)	122,000
To Customer ($/month)[2]	28.80

[1]   Average of six manufacturers; includes equipment + installation costs.

[2]   Capital, amortized at 10% for 20 years + maintenance

REFERENCES

1. Baier, J. H., Lykins, Jr., B. W., Fronk, C. A., and Kramer, S. J., "Using Reverse Osmosis to Remove Agricultural Chemicals from Groundwater, *Journal American Water Works Association*, 79(8)55-60 (1987).

2. Guerrera, A. A., "Chemical Contamination of Aquifers on Long Island, New York, *Journal American Water Works Association*, 73(4)190-199 (1981).

3. Baier, J. H. and Robbins, S. F., "Groundwater Contamination from Agricultural Chemicals", North Fork, Suffolk County, *In Proceedings ASCE National Conference on Environmental Engineering*, Boulder, Colorado, (July 6-8, 1983).

4. Baier, J. H. and Rykbsot, K. A., "The Contribution of Fertilizers to Groundwater of Long Island", *Journal National Water Well Association*, (November-December 1976).

5. DeFilippi, J. A. and Baier, J. H., "Point-of-Use and Point-of-Entry Treatment on Long Island", *Journal American Water Works Association*, 79(10):76-81 (1987).

6. Baier, J. H. "Long Island's Home Water Treatment District Experience", *In Proceedings of Water Quality Association and American Society of Agricultural Engineers Technical Papers*, Fourth Domestic Water Quality Symposium Point-of-Use Treatment and its Implications, pp. 73-82, December 16-17, 1985.

7. Wilson, K., "State Sues Herbicide Company", *Newsday*, pp. 7, 29, September 21, 1989.

8. Baier, J. H., "Small systems Opportunities For POU Industry - Local Perspective", *In Proceedings of the Water Quality Association*, San Antonio, Texas, pp. 41-44, March 14-18, 1990.

9. Bianchin, S. L., "Point-of-Use and Point-of-Entry Treatment Devices Used at Superfund Sites to Remediate Contaminated Drinking Water", *In Proceedings of Conference on Point-of-Use Treatment of Drinking Water*, Cincinnati, Ohio, Water Engineering Research Laboratory, Cincinnati, Ohio, October 6-8, 1987, EPA/600/9-88/012, (June, 1988).

10. Longley, K. E., Hanna, G. H., and Gump, B. H., "Removal of DBCP from Groundwater", Volume 1, POE/POU Treatment Devices: Institutional and Jurisdictional Factors, Water Feige Project Officer, Water Engineering Research Laboratory, Cincinnati, Ohio, (1980).

11. Rogers, K. R., "Point-of-Use Treatment of Drinking Water in San Ysidro, New Mexico", Kim R. Fox, Project Officer, Risk Reduction Engineering Laboratory, Cincinnati, Ohio, (1989).

12.   Lowry, J. D., Lowry, S. B., and Cline, J. K., Radon Removal by POE GAC Systems: Design, Performance and Cost, Kim R. Fox, Project Officer, Risk Reduction Engineering Laboratory, Cincinnati, Ohio, (1989).

13.   Lowry, J. D. and Brandow, J. E., "Removal of Rn from Water Supplies, *Journal of Environmental Engineering*, 111:4, (1985).

14.   Lowry, J. D., "Radon Progeny Accumulation in Field GAC Units", Final Report, Maine Department of Human Services, Division of Health Engineering, (March 1988).

15.   Lowry, J. D., et al., "New Developments and Considerations for Radon Removal from Water Supplies", Proceedings of the EPA Conference on Radon, November, Denver, Colorado, (1988).

16.   Kinner, N. E., Malley, Jr., J. P., Clemen, J. A., Quern, P. A., and Schell, G. S., "Radon Removal Techniques for Small Community Public Water Supplies", Kim R. Fox, Project Officer, Risk Reduction Engineering Laboratory, Cincinnati, Ohio, (1989).

17.   Bellen, G., Anderson, M., and Gottler, R., *"Point-of-Use Reduction of Volatile Halogenated Organics in Drinking Water"*, EPA/600/2-85/111, Steven Hathaway, Project Officer, Water Engineering Research Laboratory, Cincinnati, Ohio, (1985).

18.   Fox, K. R., "Field Experience with Point-of-Use Treatment Systems for Arsenic Removal", *Journal American Water Works Association*, 81(2):94-101, (February 1989).

19.   Fox, K. R. and Sorg, T. J. "Controlling Arsenic, Fluoride, and Uranium by Point-of-Use Treatment", *Journal American Water Works Association*, 79(10):81-84, (October 1987).

20.   Bellen, G., Anderson, M., and Gottler, R., *Defluoridation of Drinking Water in Small Communities*, EPA/600/2-85/110, Steven Hathaway, Project Officer, Water Engineering Research Laboratory, Cincinnati, Ohio, (1985).

CHAPTER 7

COST OF POU/POE DEVICES

INTRODUCTION

The Safe Drinking Water Act affects approximately 60,000 community water systems. Eighty-seven percent of these systems serve 3,300 or fewer people, many of which are not in compliance with the MCLs.[1] Small systems in general face severe economic constraints associated with treatment of contaminated water supplies because the unit costs of constructing and operating small central treatment systems are very high. In addition many small utilities have difficulty in attracting qualified personnel to operate small central plants.

The alternatives to treatment include development of a new well or surface water source, or connection to a neighboring public water supply. Constructing a well or surface source, or connecting to a neighboring supply may be costly. In many cases, treatment may be the only cost effective alternative.

If treatment is selected as a solution to a drinking water contamination problem, the options may be a central plant or point-of-use (POU) or point-of-entry (POE) treatment. POU/POE treatment is provided at residences or businesses to control a wide spectrum of contaminants and applications including: improving aesthetic water quality (i.e., to control taste, odor, and color), reducing levels of organic chemicals, including pesticides, control of turbidity, fluoride, iron, radium, cysts, chlorine, arsenic, nitrate, ammonia, and microorganisms. Of course, central treatment can be used to control or remove these contaminants as well.

POU/POE VS CENTRAL TREATMENT

As long as capital and operating costs can be spread over a large number of customers central treatment is cost effective but as community size decreases, per capita costs for central treatment systems increase rapidly. Economies of scale make the construction of central treatment plants for small water systems difficult.

To illustrate this effect, the relationship between monthly customer cost and average daily flow for small communities using central activated alumina treatment (for reduction of fluoride) is shown in Figure 1.[2]   As average flow decreases, the monthly customer costs increase and POU/POE treatment could become more cost effective. Costs for POU/POE treatment may be significantly lower than costs for central treatment in small communities, because no capital intensive treatment facility is required. POU/POE devices may provide a substantial cost advantage if no central treatment and distribution system exists.   However, the cost of monitoring and maintenance of these devices must be included when making a cost comparison between central treatment and POU/POE.

Small water systems frequently have difficulty hiring full time, experienced plant operators and must rely on unlicensed part time operators. This inability to retain qualified full time personnel by small utilities may result in the loss of control of finished water quality in central treatment plants.

Other comparative studies involving removal of organic compounds have also shown the effect of economics of scale.[1]   For example, assumptions were made which included 275 gallons/day/house with over 95 percent removal of dibromochloropropane (DBCP), trichloroethylene (TCE), and 1,2-dichloropropane (1,2-DCP). The costs developed for this scenario are for those central water supply systems with a distribution system already in place. Established water supply systems will already have a distribution system, thus POE is not likely to be a viable alternative except for the smallest utility or one incapable of financially building or maintaining a new central treatment plant. Cost estimates for central and POE treatment alternatives are shown in Table 1.

Requiring treatment of well water to combat the contamination of individual wells from leaking underground storage tanks, municipal landfills, and agricultural chemicals is one that state and local governments will have to face increasingly over time. Trailer parks and new subdivisions are other entities that may have to consider treatment to meet new MCLs. It is these situations where decisions will have to be made as to whether it is feasible to connect these homeowners to central treatment, install central treatment and a distribution network, or provide POE units. Connecting to an existing central supply is usually the first alternative considered.

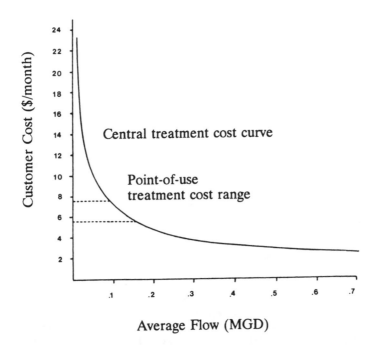

Figure 1. Central and Point-of-Use Treatment Costs for Activated Alumina

TABLE 1. POE VS. CENTRAL SYSTEM COST[a]

Households	Contaminant	Initial Concentration[b] µg/L	Central System Cost ¢/1000 gal	Average POE Cost ¢/1000 gal
10	DBCP	50	1385	475
25	DBCP	50	669	475
50	DBCP	50	398	475
10	TCE	100	1395	675
25	TCE	100	679	675
50	TCE	100	408	675
10	1,2-DCP	100	1494	800
25	1,2-DCP	100	750	800
50	1,2-DCP	100	465	800

[a] Distribution system already in place.
[b] greater than 95 percent removal.

In this next analysis, shown below, is a trailer park and a sub-division which need drinking water treatment to remove first, an organic contaminant (trichloroethylene) and second, an inorganic contaminant (nitrate). Each scenario will compare central treatment with distribution system costs versus POE installation. Each residential area has 150 homes (approximately 500 consumers) requiring about 40 gpm total. The trailer park being very densely populated requires 3,400 feet of pipe whereas the subdivision requires 15,840 feet (3 miles). Eight inch PVC pipe is used for cost estimating incorporating additional costs for trenching, embedment, backfill, paving, and variable connection costs for different population densities.[3]

GAC Analysis

Trichloroethylene (TCE) is one of the most common contaminants in groundwater. GAC can be used to remove TCE. Central system GAC updated cost assumptions include: an empty bed contact time of 10 minutes, a carbon service life of 165 days, 30 percent excess capacity, and 10 percent financing for 20 years. The POE unit consists of two adsorbers in-series, each with 2 cubic feet of F-400 carbon, 4.1 minutes empty bed contact time, loading rate of 4 gpm/square foot, and 10 percent financing for 10 years. The GAC POE capital cost is 2,000 dollars with an annual carbon replacement cost of one tank per year at 420 dollars with a 15 dollar per month maintenance charge. An influent level of 100 µg/L of TCE is being treated to 5 μg/L (the MCL) in each case.   If packed tower aeration is used prior to GAC, the GAC bed life can be extended by as much as 80 percent by removing most of the TCE prior to GAC adsorption.   This will result in an over-all reduction in total cost although the aerator capital cost will be approximately $3,000 and will require more electricity.

For central treatment, another alternative is considered which incorporates four smaller GAC units of 10 gpm each rather than one unit of 40 gpm.   In some circumstances, this may save on the amount of pipe needed given population clusters.   In this case, it was assumed that 25 percent less pipe was needed.   As can be seen in Table 2, central treatment for the densely populated trailer park is the least expensive scenario.   However, the central treatment subdivision costs are within 10 percent of the POE cost.   Distribution system costs account for about 70 percent of the total costs for the subdivision and only 50 percent of the trailer park's cost.   Should ductile iron pipe be used instead of PVC, distribution costs would double, thus making POE cost-effective for even more homes.

TABLE 2.   GAC COST SCENARIOS FOR TCE REMOVAL

| Residential Area | CENTRAL TREATMENT[a] | | POE |
	1 GAC Unit (40 gpm)	4 GAC Units (10 gpm each)	150 GAC POE Units
Trailer Park			
(house/year)	$ 357	$ 636	$ 690
(1000 gals)	$3.70	$6.60	$7.16
Subdivision			
(house/year)	$ 619	$ 837	$ 690
(1000 gals)	$6.42	$8.68	$7.16

(a)   includes cost for distribution system

The central treatment scenario incorporating four 10 gpm units proved to be very costly. The 25 percent reduction in pipe was not enough to offset the extra treatment device costs.

Ion Exchange Analysis

In order to remove nitrate below the 10 mg/L standard (95% removal), ion exchange can be used.   Nitrate contamination of drinking water supplies has been increasing over the years mainly because of normal applications of agricultural fertilizers leaching into groundwater contaminating not only rural wells, but wells on the fringe of some very large cities. Ion exchange central treatment cost include: daily regeneration, 25 cubic feet of resin, 4.7 minute empty bed contact time, with 10 percent financing for 20 years.   Ion exchange POE assumptions include: 2,000 dollars purchase price, auto-regeneration, 15 dollars/ month service contract, with 10 percent financing for 10 years. Table 3 displays the cost comparing ion exchange central treatment versus POE. The four unit scenario is not included since the costs were so prohibitive in the GAC example.

Once again, the trailer park is least expensive for the central treatment. However, because of the lower POE cost for ion exchange versus GAC, the difference is not as large. The subdivision scenario shows central treatment to be approximately 20 percent more expensive than installing 150 POE units to remove nitrate.

TABLE 3.   ION EXCHANGE COST SCENARIOS FOR NITRATE REMOVAL

Residential Area	CENTRAL TREATMENT[a] 1 Ion Exchange Unit 40 gpm	POE 150 Ion Exchange POE Units
Trailer Park (house/year) (1000 gals)	$ 312 $3.24	$ 480 $4.98
Subdivision (house/year) (1000 gals)	$ 574 $5.96	$ 480 $4.98

(a)   includes cost for distribution system

In a draft document prepared by the U.S. Army for their drinking water installations, six processes relative to treatment of chronic health effects contaminants were cited. These consisted of reverse osmosis, cation exchange, anion exchange, activated alumina, direct filtration, and GAC adsorption. Three treatment scenarios were considered: (1) treatment of drinking water only (separate faucet), (2) all the cold water at a single faucet, and (3) water supply for the entire structure.[4] Drinking water treatment capacities were approximately 730, 1,000 to 40,000 and over 100,000 gallons per year, respectively. The cost evaluations for both POU systems and central treatment included capital costs (equipment and installation) and operation/maintenance costs (power, chemicals, labor, and replacement items for POU such as cartridges and filters). Treatment of drinking water with a separate faucet was lower in cost than central treatment in all sizes of communities evaluated for reverse osmosis, GAC, cation exchange, and direct filtration. When treating water for the entire structure, POE treatment was generally less in cost than centralized treatment for communities of less than 100 residences.

OTHER COST STUDIES

As one might expect, the cost of POU/POE devices vary. Typical initial costs for POU activated carbon devices reported by Ebbert in 1985 ranged from $30 for flow through units to $300 for under-the-sink units with a separate faucet.[5] Replacement filter costs ranged from $4 to $60 depending on the type of treatment device. For POU/POE reverse osmosis units, the initial costs ranged from $450 to $850 depending on the membrane area. Membrane replacement costs varied from $70 to $140.

The average capital cost for granular activated carbon POU/POE treatment in Silverdale, Pennsylvania was $290. The total average monthly cost to the customer was $5.60 which included $2.77 for maintenance and $2.83 for capital cost based on an amortization of 10% for 20 years.[6]

The average capital cost for GAC POU/POE treatment in Rockaway Township, New Jersey was $300 (1984). The total estimated average replacement costs were $2.12 per month based on an assumed GAC cartridge life of two years and a yearly replacement of particulate filters. The total average cost to the customer was $5.06 which included $2.12 for maintenance and $2.94 for capital cost based on an amortization of 10% for 20 years.[6]

In 1989, information was requested from POU/POE manufacturers, suppliers, and regulators on: (1) the types of devices that are currently being used, (2) what contaminants are being removed by these devices, (3) the effectiveness of these devices, and (4) available data on the cost of the devices.[7] A total of 384 letter requests were sent and 164 responses were received. All of the data was computerized and a breakdown of the POU and POE technology capital costs are shown in Tables 4 and 5. These tables represent a range of costs primarily for the basic units and tend to be somewhat less than the costs presented previously. The manufacturers' literature did not usually list cost of installation or other needed equipment.

In a study done for Suffolk County, New York representative costs for POU/POE systems were determined.[8] These costs are presented in Table 6. Many homes in Suffolk County are using POU units to remove organic chemicals from contaminated wells. These units are generally installed and maintained by the homeowner. However, to more adequately protect the users of these units, Suffolk County is evaluating centralized monitoring control. The total costs of this approach for two towns in Suffolk County are shown in Table 7.

TABLE 4.   POU DATABASE TECHNOLOGY COST BREAKDOWN[7]

TECHNOLOGY	MINIMUM COST	MAXIMUM COST	AVERAGE COST
DESCALER	280.00	280.00	280.00
DISTILLATION	214.38	1749.00	817.38
FILTRATION	13.75	899.00	258.63
GAC	4.54	822.25	136.62
ION EXCHANGE	195.00	275.00	235.00
RO	39.95	999.00	353.10
UV	254.00	732.00	550.67
COMBINATION *	87.50	6200.00	1559.08

*   Any of the above technologies in series (e.g. filtration/GAC/RO, etc.)

TABLE 5.   POE DATABASE TECHNOLOGY COST BREAKDOWN[7]

TECHNOLOGY	MINIMUM COST	MAXIMUM COST	AVERAGE COST
AERATION	1650.00	1650.00	1650.00
CHLORINE	235.85	246.95	241.40
DISTILLATION	640.43	640.43	640.43
FILTRATION	48.75	852.20	359.22
GAC	539.00	1329.85	939.71
ION EXCHANGE	415.00	1250.00	956.67
NEUTRALIZATION	335.00	395.00	368.33
RO	79.00	6340.00	2996.02
SOFTENING	425.00	1200.00	731.67
UV	317.00	637.00	486.00
COMBINATION *	379.00	1650.00	750.00

*   Any of the above technologies in series (e.g. filtration/GAC/RO, etc.)

TABLE 6.   REPRESENTATIVE COSTS FOR POU/POE UNITS
(1985 DOLLARS)[8]

UNIT TYPE	SINGLE TAP (POU)		WHOLE HOUSE (POE)	
	COST RANGE ($)	INSTALLATION COSTS ($)	COST RANGE ($)	INSTALLATION COSTS ($)
REVERSE OSMOSIS	500-800	70-150	6,000-8,500*	250-350
GRANULAR ACTIVATED CARBON	200-350	60-100	1,100-3,000	75-150
DISTILLATION	200-800	100-150	9,500-11,000	200-300

*   This includes the RO unit, storage tank, and a dispenser pump

TABLE 7.   ESTIMATED COSTS (IN 1985 DOLLARS) FOR CENTRALIZED
MANAGEMENT OF POU SYSTEMS. SUFFOLK, COUNTY, NEW YORK[8]

TOWN	TOTAL CAPITAL COST ($)[a]	ANNUAL OPERATING COST ($)[b]	TOTAL ANNUAL COSTS ($)[c]
RIVERHEAD	1,400,000[d]	80,000	360,000
SOUTHHOLD	1,800,000[e]	110,000	470,000

a   Includes a contingency of 15% and associated costs of 5%.

b   Assumes $50.00 medium replacement costs for individual units.

c   Yearly operating costs + amortized capital cost based on 12% interest over an eight year period.

d   Based on 1550 single tap units at $600.00 each and 100 whole house units at $2000.00 each, including installation.

e   Based on 2150 single tap units at $600.00 each and 100 whole house units at $2000.00 each, including installation.

An EPA in-house study in 1988 compared the cost per gallon of four types of POU/POE systems.[9]   Manufacturers were contacted for current costs and some assumptions were made to produce the following analysis. For POU treatment costs:

GAC
(Single Tap)

- Capital Cost   Installation Cost   Total Capital Cost
  $275   + $ 80   = $355

- $/year O&M cost = $50

- 15% Contingency + 5% Associated Costs = 1.2

- Amortization Factor = 12% / 8 years = 0.2013

- Total Capital Cost = $355 X 1.2 X 0.2013 = $86/year

- Total Annual Cost = $86 + $50 = $136/year

- $/1,000 gal = $136/2X10^{-5}/3,650/100 = $18.63/1,000 gal
                                           = $ 1.9 ¢/gallon

Reverse
Osmosis
(Single
 Tap)

- Capital Cost   Installation Cost   Total Capital Cost
  $650   + $100   = $760

- $760 X 1.2 X 0.2013 = $184/year

- Total Annual Cost = $184 + $50 = $234/year

- $/1,000 gal = $234/2X10^{-5}/3,650/100 = $32.05/1,000 gal
                                           = $ 3.2 ¢/gallon

Ion
Exchange
(Single
Tap)

- Capital Cost   Installation Cost   Total Capital Cost
  $200   + $ 80   = $280

- $280 X 1.2 X 0.2013 = $ 68/year

- Total Annual Cost = $ 68 + $50 = $118/year

- $/1,000 gal = $118/2X10^{-5}/3,650/100 = $16.16/1,000 gal
                                           = $ 1.6 ¢/gallon

Distillation
(Single Tap)

- Capital Cost   Installation Cost   Total Capital Cost
  $500   + $125   = $625

- $625 X 1.2 X 0.2013 = $151/year

- Total Annual Cost = $151 + $50 = $201/year

- $/1,000 gal = $201/2X10^{-5}/3,650/100 = $27.53/1,000 gal
                                           = $ 2.8 ¢/gallon

GAC
(Whole House)

- Capital Cost   Installation Cost   Total Capital Cost
  $2,050   + $115   = $2,165

- $2,165 X 1.2 X 0.2013 = $523/year

- Total Annual Cost = $523 + $50 = $573/year

- $/1,000 gal = $573/2.8X10^{-4}/3,650/100 = $ 5.12/1,000 gal
  = $ 0.5 ¢/gallon

Reverse
Osmosis
(Whole
House)

- Capital Cost   Installation Cost   Total Capital Cost
  $7,250   + $300   = $7,550

- $7,550 X 1.2 X 0.2013 = $1,824/year

- Total Annual Cost = $1,824 + $50 = $1,874/year

- $/1,000 gal = $1,874/2.8X10^{-4}/3,650/100 = $18.34/1,000
  gal   = $ 1.8 ¢/gallon

Ion
Exchange
(Whole
House)

- Capital Cost   Installation Cost   Total Capital Cost
  $1,750   + $175   = $1,925

- $1,925 X 1.2 X 0.2013 = $465/year

- Total Annual Cost = $465 + $50 = $515/year

- $/1,000 gal = $515/2.8X10^{-4}/3,650/100 = $ 5.04/1,000 gal
  = $ 0.5 ¢/gallon

Distillation
(Whole House)

- Capital Cost   Installation Cost   Total Capital Cost
  $10,250   + $250   = $10,500

- $10,500 X 1.2 X 0.2013 = $2,536/year

- Total Annual Cost = $2,536 + $50 = $2,586/year

- $/1,000 gal = $2,586/2.8X10^{-4}/3,650/100 = $25.30/1,000
  gal   = $ 2.5 ¢/gallon

Capital and operating costs for individual unit processes are given in the following section based on 1990 costs.

COST OF POU/POE DEVICES

EPA has compiled estimates of the capital and operating cost for various types of POU/POE devices.[10]   The capital cost includes installation under the kitchen sink, any prefilters required, initial charge of media, equipment housing, a special faucet, required tubing, fittings and adapters. The meter box construction cost includes the items listed above and also include the cost of PVC piping from the meter box to the house ($20 for pipe and $45 for installation).

The operating and maintenance costs include labor and materials but no electricity costs. Labor costs are for collecting water samples for testing the units performance for replacing media, filters, and membranes. Maintenance material requirements are for sampling, filter, membrane, and media replacement. Included in materials cost is the cost of laboratory testing and repairs for tubing or fittings.

Granular Activated Carbon

Construction Costs

Construction costs for POU/POE GAC contactors are presented in Table 8 for installations located under the kitchen sink. The cost for the contactor is based on using either a quality plastic or stainless steel housing and a replaceable carbon cartridge. The housing is 25 to 30 cm (10 to 12 in) long and 10 to 13 cm (4 to 5 in) in diameter.

Carbon quantities contained in the GAC contactor range from 0.002 to 0.003 m^3 (0.06 to 0.09 cu ft). The quantity of water which can be treated by the GAC filter unit, assuming the application is organics removal, will depend upon the concentration and type of organics in the untreated water. However, a reasonable range in treated water capacity is between 3,800 to 11,300 L (1,000 to 3,000 gal), based on the above carbon volumes.

TABLE 8. CONSTRUCTION COST SUMMARY FOR GRANULAR
ACTIVATED CARBON POU TREATMENT[10]

Cost Category	Construction Cost, $ Installation Located Under Sink or in Basement
GAC Contactor	$155
Faucet, Copper or Plastic Tubing, and Fittings	45
Labor, Installation	45
Subtotal	$245
Contingency	35
Total	$280

NOTE: 1. For an additional cost of $55, a bypass line and needle valve can be installed to blend treated and untreated water to achieve an acceptable contaminant level in the final product water.

2. For an additional cost of $45, a water meter can be installed after the GAC contactor to record the amount of water treated and to help the utility estimate the GAC contactor life.

Operation and Maintenance Requirements and Costs

Operation and maintenance requirements for point-of-use GAC contactors are only for labor and material costs. No electricity is required since the water pressure delivered to the home is sufficient for operation of the GAC contactor. The operation and maintenance requirements are separated into three cost categories: sampling/testing, carbon cartridge replacement, and repairs.

A summary of operation and maintenance requirements is shown in Table 9. A range of requirements is shown for each cost category so that the level of sampling, frequency of carbon replacement, and level of repairs can be selected by the user of this cost information.

Labor requirements are for collecting samples of contactor product water, for replacement of the GAC contactor cartridge, and for infrequent repairs of piping or contactor leaks. The major sample/testing would consist of testing for organics or a specific heavy metal contaminant such as mercury. When the organics or mercury contaminant level rises above the allowable limit, it is time to replace the carbon cartridge. Another type of testing done occasionally on GAC contactors is bacterial testing consisting of standard plate counts and total coliform tests. If the total coliform test is positive, a fecal coliform test should be conducted.

TABLE 9.   OPERATION AND MAINTENANCE SUMMARY FOR GRANULAR ACTIVATED CARBON POU TREATMENT

Requirements	Labor,(a) hr/yr	Materials,(b) $/yr	Total Cost, $/YR Specific Condition	Total Cost, $/YR Mid-Range Condition
Sampling/Testing Frequency				
2/yr	2	65	89	
4/yr	4	135	183	$183
6/yr	6	200	272	
Carbon Replacement Frequency every				
1 yr	1	10	22	
2 yrs	0.5	5	11	11
3 yrs	0.4	3	6	
Repairs				
Low	1	10	22	
Average	2	20	44	44
High	3	35	71	

Total Cost for Mid-Range Condition = $238/yr

NOTE: (a)   Total cost is based on $12.00/hour of labor.
      (b)   Material costs for sampling/testing represents the cost for laboratory testing. Material costs for regeneration assume that the media is regenerated locally. If extensive shipping costs are incurred, the regeneration materials cost will have to be increased to account for this shipping.

Maintenance material requirements are for sampling, GAC contactor cartridge replacement, and for other repair components as required. Material requirements for sampling are the costs of the laboratory analyses required for organics or inorganic tests and bacterial counts. The material cost for repairs is for infrequent replacement of a leaking piece of pipe or fitting, or the carbon contactor housing.

Reverse Osmosis

Construction Costs

Construction costs for POU/POE reverse osmosis systems are presented in Table 10 for installations located under the kitchen sink. The cost for manufactured equipment includes a reverse osmosis membrane unit, a pre-filter, a pressure reservoir, a small GAC post-contactor, a special faucet and all required tubing, fittings, and adapters. In virtually all commercially available equipment, cellulose acetate membranes are utilized. The disadvantage of using thin film composite membranes is the necessity to use GAC pre-treatment for chlorine removal. This GAC must be maintained in an active state, for any penetration of chlorine will result in membrane deterioration.

The 5-micron pre-filter removes particles such as sediment or dirt and is necessary to protect the reverse osmosis membrane. The reverse osmosis membrane unit has a capacity of producing up to 19 L (5 gal) of water per day.

TABLE 10.   CONSTRUCTION COST SUMMARY FOR REVERSE OSMOSIS POU TREATMENT

Cost Category		Construction Cost, $
Manufactured Equipment		$330
Labor, Installation		60
	Subtotal	$390
Contingency		70
	Total	$460

NOTE: For an additional cost of $45, a water meter can be installed after the treatment unit to record the amount of water treated and to help the utility estimate membrane life.

A pressure reservoir stores treated water in a plastic lined steel chamber with a 12 L (3.2 gal) storage capacity; the reservoir is pressurized by air behind a rubber diaphragm. The GAC contactor is located in the line between the storage reservoir and the faucet and is 5 cm (2 in) in diameter by 18 cm (7 in) long; the GAC contactor removes chlorine and taste and odor causing compounds which may pass through the reverse osmosis membrane.

Operation and Maintenance Requirements and Costs

Operation and maintenance requirements for POU/POE reverse osmosis systems are only for labor and material costs. No electricity is required since the water pressure typically delivered to the home (40 psi or greater) is sufficient for operation of the reverse osmosis unit. The operation and maintenance requirements are separated into four cost categories: sampling/ testing, pre-filter and GAC contactor replacement, reverse osmosis membrane replacement, and repairs.

A summary of operation and maintenance requirements is shown in Table 11. A range of requirements is shown for each cost category, so that the level of sampling frequency for pre-filter and GAC contactor replacement, interval of reverse osmosis membrane replacement, and level of repairs can be selected by the user of this cost information.

Labor requirements are for collecting water quality samples, testing reverse osmosis system performance, replacing the pre-filter and GAC contactor, replacing the reverse osmosis membrane module, and for infrequent repairs on the system. Sampling/testing requirements are for the contaminant(s) the system is designed to remove; when the contaminant level rises above the MCL, it is time for replacement of the reverse osmosis membrane. The pre-filter and GAC contactor typically have a shorter useful life than a reverse osmosis membrane and are replaced more frequently.

Maintenance material requirements are for sampling, filter and GAC replacement, reverse osmosis and membrane replacement, and any other repair components required. Material requirements for sampling are for the cost of laboratory analyses required to test the level of the contaminant(s) being removed. Material requirements for pre-filter and GAC contactor replacement are the costs for new pre-filter cartridges and a new GAC contactor. The material cost for the reverse osmosis unit is for replacement of the reverse osmosis membrane module which fits inside the membrane unit. Material requirements for repairs are for tubing or fittings.

TABLE 11. OPERATION AND MAINTENANCE SUMMARY FOR REVERSE OSMOSIS POU TREATMENT

Requirements	Labor,[a] hr/yr	Materials,[b] $/yr	Total Cost, $/YR Specific Condition	Mid-Range Condition
Sampling/Testing				
Frequency				
2/yr	2	45	69	
4/yr	4	90	138	$138
6/yr	6	135	207	
Pre-filter and GAC Contactor Replacement				
Frequency				
1/yr	1	20	32	
2/yr	2	45	69	69
Reverse Osmosis Membrane Replacement				
Interval every				
1 yr	1	85	97	
2 yrs	0.5	40	46	46
3 yrs	0.4	30	33	
Repairs				
Low	1	10	22	
Average	2	20	44	44
High	3	35	71	

Total Cost for Mid-Range Condition = $297/yr

NOTE: (a) Total cost is based on $12.00/hour of labor.
(b) Material costs for sampling/testing represents the cost for laboratory testing.

Cation Exchange

Construction Costs

Construction costs for POU/POE cation exchange systems are presented in Table 12. The costs include a cation exchange canister and resin, a special faucet, and all required tubing, fittings, and adapters. The fiberglass cation exchange canister is about 15 cm (6 in) in diameter by 58 cm (23 in) long, and contains an effective resin volume of 0.008 m^3 (0.3 cu ft). The quantity of water which can be treated by a cation exchange unit prior to regeneration depends on the chemical composition of the water being treated, and the contaminants which are being removed. Assuming the following: a hardness concentration of 600 mg/L as $CaCO_3$; a resin exchange capacity of 144 kg of hardness as $CaCO_3$ per m^3 of resin (9 lb/cu ft of resin); a resin volume of 0.008 m^3 (0.3 cu ft) and a water use of 7.6 L (2 gal) per day; the unit could operate for 270 days prior to regeneration. For the same conditions except a water hardness of 200 mg/L as $CaCO_3$, the unit could operate for 810 days before regeneration. Pilot-scale testing should be performed to determine the actual regeneration frequency for a specific water supply. Many POE units could be installed underground at the property line in certain climates but local ordinances should be checked.

Operation and Maintenance Requirements and Costs

Operations and maintenance requirements for point-of-use cation exchange units are only for labor and material costs. No electricity is required since the water pressure delivered to the home is sufficient for operation of the cation exchange unit. The operation and maintenance requirements are separated into three cost categories: sampling/testing, resin regeneration, and repairs.

A summary of operation and maintenance requirements is shown in Table 13. A range of requirements is shown for each cost category, so that the level of sampling, frequency of resin regeneration, and level of repairs can be selected by the user of this cost information.

Labor requirements are for collecting samples on cation exchange performance, replacement of exhausted resin with regenerated resin, off-site regeneration of exhausted resin or replacement of resin, and for infrequent repair of leaks in piping, fittings, or in the cation exchange unit. The major sampling/testing effort is to run tests for contaminants. When the contaminant concentration exceeds the MCL, the exhausted canister should be replaced. Another type of testing done occasionally is bacterial testing consisting of standard plate counts and total coliform tests. If the total coliform test is positive, a fecal coliform test should be conducted.

TABLE 12.   CONSTRUCTION COST SUMMARY FOR CATION EXCHANGE
POU/POE TREATMENT

Cost Category	Construction Cost, $	
	Installation Located Under Sink or in Basement	Installation Located Underground at the Property Line in a Meter Box
Cation Exchange Canister and Resin	$200	$200
Plastic Meter Box and 10 in PVC Pipe Collar	---	35
PVC Pipe to House	---	20
Faucet, Copper or Plastic Tubing, and Fittings	45	45
Labor, Installation	45	135
Subtotal	$290	$435
Contingency	45	65
Total	$335	$500

NOTE: 1.   For an additional cost of $55, a bypass line and needle valve can be installed to blend treated and untreated water to achieve an acceptable contaminant level in the final product water.

2.   For an additional cost of $45, a water meter can be installed after the treatment unit to record the amount of water treated and to help the utility estimate the GAC contactor life.

3.   For property line installations, costs for PVC pipe to the house ($20) and for installation of this line ($45) may be low for extensively landscaped yards.

Maintenance material requirements are for sampling, resin regeneration, and repairs. Material requirements for sampling are the costs of the laboratory analysis. The material cost for repairs is for infrequent replacement of a leaking piece of pipe or fitting or the cation exchange unit housing.

TABLE 13.   OPERATION AND MAINTENANCE SUMMARY FOR CATION EXCHANGE POU TREATMENT

Requirements	Labor, (a) hr/yr	Materials, (b) $/yr	Total Cost, $/YR Specific Condition	Total Cost, $/YR Mid-Range Condition
Sampling/Testing Frequency				
2/yr	2	45	69	
4/yr	4	90	138	$138
6/yr	6	135	207	
Resin Regeneration Frequency every				
1 yr	1	35	47	
2 yrs	0.5	15	21	21
3 yrs	0.4	10	13	
Repairs				
Low	1	10	22	
Average	2	20	44	44
High	3	35	71	

Total Cost for Mid-Range Condition   = $203/yr

NOTE: (b) Total cost is based on $12.00/hour of labor.
      (a) Material costs for sampling/testing represents the cost for laboratory testing.

Activated Alumina

Construction Costs

Construction costs for the POU/POE defluoridation by activated alumina filtration are presented in Table 14 both for installations located under the sink (or in the basement) and for installations located underground at the property line. For either location, the activated alumina filter canister is the same, a 14.2 L (0.5 cu ft) fiberglass cylinder which is 18 cm diameter by 56 cm high (7 in diameter by 22 in high).

TABLE 14.   CONSTRUCTION COST SUMMARY FOR ACTIVATED ALUMINA
POU TREATMENT

	Construction Cost, $
Cost Category	Installation Located Under Sink or in Basement
Activated Alumina Filter Canister	$225
Faucet, Copper or Plastic Tubing, and Fittings	45
Labor, Installation	45
Subtotal	$315
Contingency	45
Total	$360

NOTES: 1.   For an additional cost of $55, a bypass line and needle valve can be installed to blend treated and untreated water to achieve an acceptable contaminant level in the final product water.

2.   For an additional cost of $45, a water meter can be installed after the treatment unit to record the amount of water treated and to help the utility estimate the unit life.

Operation and Maintenance Requirements

Operation and maintenance requirements for POU/POE activated alumina fluoride removal are only for labor and material costs only. Since the water pressure delivered to the home is sufficient for operation of the activated alumina filter, no electricity is required. The operation and maintenance requirements are separated into three cost categories: sampling/testing, media regeneration, and repairs.

A summary of operation and maintenance requirements is shown in Table 15. A range of requirements is shown for each cost category so that the level of sampling, frequency of regeneration, and level of repairs can be selected by the user of this cost information.

Labor requirements are for collecting samples of product water for fluoride analysis, removing the filter for regeneration or replacement of activated alumina media, and for infrequent repairs of piping or filter leaks. The major sample/testing is to run fluoride tests on the product water. When the fluoride level rises above the MCL, the treatment canister should be replaced with one containing regenerated or new activated alumina. Other types of testing which are occasionally performed on POU/POE units are standard plate counts and total and fecal coliform testing.

Maintenance material requirements are for sampling, resin regeneration/replacement, and repair. Material requirements for sampling are for the costs of laboratory analyses for fluoride and bacterial counts. The costs are fairly low because it is assumed that a number of homes would be tested at the same time, lowering the per unit cost. The material requirement for regeneration is actually for the entire cost of having the media regenerated or replaced. The material cost for repairs is for infrequent replacement of a leaking piece of pipe or fitting or the fiberglass filter housing.

Anion Exchange

Construction Costs

Construction costs for POU/POE anion exchange units are presented in Table 16. The cost of the anion exchange unit is based on using a fiberglass housing and a replaceable anion exchange resin cartridge. The housing is 16 cm (6 1/4 in) in diameter by 58 cm (23 in) long and has a resin capacity of 0.008 m^3 (0.3 cu ft). The quantity of water which can be treated by the anion exchange unit depends on the contaminant that is being removed. For nitrate removal applications, the quantity treated is a function of nitrate, sulfate, and total anion concentrations. The anion exchange resin preferentially removes sulfate over nitrate; therefore when the exchange sites are filled, the unit will begin to release nitrate, and an increase in the nitrate concentration will occur in the effluent from the unit. For a nitrate concentration of 100 mg/L as nitrate, or 23 mg/L

TABLE 15.   OPERATION AND MAINTENANCE SUMMARY FOR ACTIVATED ALUMINA POU TREATMENT

Requirements	Labor,(a) hr/yr	Materials,(b) $/yr	Total Cost, $/YR	
			Specific Condition	Mid-Range Condition
Sampling/Testing Frequency				
2/yr	2	20	44	
4/yr	6	45	117	$117
6/yr	12	70	214	
Media Regeneration Frequency every				
1 yr	2	110	134	
2 yrs	1	55	67	67
4 yrs	0.5	30	36	
Repairs				
Low	1	10	22	
Average	2	20	44	44
High	3	35	71	

Total Cost for Mid-Range Condition   = $228/yr

NOTE: (a) Total cost is based on $12.00/hour of labor.
       (b) Material costs for sampling/testing represents the cost for laboratory testing. Material costs for regeneration assume that the media is regenerated locally. If extensive shipping costs are incurred, the regeneration materials cost will have to be increased to account for this shipping.

TABLE 16.   CONSTRUCTION COST SUMMARY FOR ANION EXCHANGE
POU TREATMENT

Cost Category	Construction Cost, $ Installation Located Under Sink or in Basement
Anion Exchange Canister and Resin	$300
Faucet, Copper or Plastic Tubing, and Fittings	45
Labor, Installation	45
Subtotal	$390
Contingency	55
Total	$445

NOTE: 1.   For an additional cost of $55, a bypass line and needle valve can be installed to blend treated and untreated water to achieve an acceptable contaminant level in the final product water.
2.   For an additional cost of $45, a water meter can be installed after the treatment unit to record the amount of water treated and to help the utility estimate the GAC contactor life.

as N, and a 100 mg/L sulfate concentration, the nitrate exchange capacity is 15.3 kg/m^3 (6,700 grains per cu ft) of exchange resin. For an anion exchange resin volume of 0.008 m^3 (0.3 cu ft) the nitrogen removal capacity is 0.13 kg (0.29 lb) of nitrogen. At a water usage of 7.5 L (2 gal) per day the total nitrogen feed rate is 1.7×10^{-4} kg/d (3.8×10^{-4} lb/day) of nitrogen, resulting in a resin life of 763 days, or about three years. During this time period, the anion exchange resin removes both sulfate and nitrate from the feed water.

Operation and Maintenance Requirements and Costs

Operation and maintenance requirements for POU/POE anion exchange units are only for labor and material costs. No electricity is required since the water pressure delivered to the home is sufficient for operation of the anion exchange unit. The operation and maintenance requirements are separated into three cost categories: sampling/testing, anion exchange resin regeneration, and repairs.

Labor requirements are for collecting samples on the anion exchange unit product water, replacement of the anion exchange resin, and for infrequent repairs of leaks in the piping or the anion exchange unit. The major sample/testing is for nitrate concentration. When the nitrate concentration begins to approach the MCL, it is time to replace the anion exchange resin. Another type of testing done occasionally on anion exchange units is bacterial testing consisting of standard plate counts and total coliform tests. If a total coliform test is positive, a fecal coliform test is conducted.

Maintenance material requirements are for sample collection, replacement resin, and miscellaneous repair parts. For resin, either complete replacement or regeneration should be used, whichever is the more cost-effective approach.

Material requirements for sampling are the costs of the laboratory analyses required for nitrate tests and bacterial counts. The material costs for repairs is for infrequent replacement of a leaking piece of pipe or fitting or the anion exchange unit housing.

A summary of operation and maintenance requirements is shown in Table 17. A range of requirements is shown for each cost category, so that the level of sampling, frequency of resin regeneration, and level of repairs can be selected by the user of this cost information.

TABLE 17.   OPERATION AND MAINTENANCE SUMMARY FOR ANION EXCHANGE POU TREATMENT

Requirements	Labor, (a) hr/yr	Materials, (b) $/yr	Total Cost, $/YR Specific Condition	Total Cost, $/YR Mid-Range Condition
Sampling/Testing Frequency				
2/yr	2	45	69	
4/yr	4	90	138	$138
6/yr	6	135	207	
Resin Regeneration Frequency every				
1 yr	1	65	77	
2 yrs	0.5	35	41	41
3 yrs	0.4	20	23	
Repairs				
Low	1	10	22	
Average	2	20	44	44
High	3	35	71	

Total Cost for Mid-Range Condition   = $223/yr

NOTE: (a) Total cost is based on $12.00/hour of labor.
      (b) Material costs for sampling/testing represents the cost for laboratory testing. Material costs for regeneration assume that the media is regenerated locally. If extensive shipping costs are incurred, the regeneration materials cost will have to be increased to account for this shipping.

REFERENCES

1. Goodrich, J.A., Lykins, B.W., Jr., Adams, J.Q., Clark, R.M., "Safe Drinking Water For The Little Guy: Options and Alternatives," in 1990 Annual Conference Proceedings, American Water Works Association, Cincinnati, OH, p. 1111, June 17-21, 1990.

2. Bellen, G., Anderson, M., Gottler, R., Management of Point-Of-Use Drinking Water Treatment System, Water Engineering Research Laboratory, Office of Research and Development, U.S. Environmental Protection Agency, Cincinnati, OH 45268, p. 13, (July 1986).

3. HDR, "Standardized Cost For Water Distribution Systems", Draft, Drinking Water Research Division, Cincinnati, Ohio, May 1990.

4. Water Quality Information Paper, "POU Technology for Water Quality Improvement", (Draft), U.S. Army Environmental Hygiene Agency, Aberdeen Proving Ground, Maryland, (1987).

5. Ebbert, S., Kern, M., Lobst, J., Moyer, R. A., and Wohlbach, H., "A Homeowners Guide to Safer Drinking Water", (Emmaus, PA: Rodale Press, Inc., 1985).

6. Anderson, M.R., Gottler, R.A., and Bellen, G.E., "Point-of-Use Treatment Technology to Control Organic and Inorganic Contaminants," Part II, *Water Technology*, 41-47, (October 1984).

7. Lykins, Jr., B. W., Goodrich, J. A., and Clark, R. M., "POU/POE Devices: Availability, Performance, and Cost", Proceedings of 1989 ASCE National Conference on Environmental Engineering, Austin, Texas, July 10-12, 1989.

8. "Point-of-Use Water Supply Treatment Systems", Riverhead and Southhold, Suffolk County, New York (Boston, MA: ERM-Northeast, Inc., 1986).

9. Eilers, R., "In-House Cost Summary on POU/POE", U.S. EPA, Drinking Water Research Division, Cincinnati, Ohio, (1988).

10. Gumerman, R.C., Culp, R.L. and Hansen, S.P., "Estimating Water Treatment Costs", EPA-600/2-79-162 a,b,c,d, Drinking Water Research Division, Cincinnati, Ohio, August, (1979).

CHAPTER 8

STATE AND FEDERAL REGULATIONS OF POU/POE

INTRODUCTION

There has been a steady increase in the purchase and use of POU/POE devices in the United States. Along with this growth has come increasing oversight and regulation at the Sate and Federal levels. Thirteen states are considering legislation that effects the industry.[1] Many states are still in the process of reviewing and enacting legislation and ten states have laws in place that effect the industry. In addition, Federal guidelines and regulations have been proposed and promulgated by the Federal Trade Commission, the U.S. Environmental Protection Agency, the U.S. Army Corps of Engineers, and the Department of Housing and Urban Development. The Canadian Government has also established legislation controlling POU/POE devices. These activities are discussed in the following sections.

THE U.S. POSITION

As mentioned previously, the U.S. Federal Government has several regulatory activities that affect the POU/POE industry. These activities are discussed in the following sections.

USEPA, Office Of Pesticides And Toxic Substances

The Office of Pesticides and Toxic Substances of THE USEPA has limited authority to regulate POU/POE devices under the Federal Insecticide, Fungicide, and Rodenticide Act (FIFRA).[2] These devices have the following characteristics:

1) units that consist of only a physical or mechanical means of preventing, destroying, repelling, or mitigating any microorganisms or pests (e.g. devices); and

2) units that incorporate a chemical antimicrobial agent or units that consist of a combination of physical and chemical treatment intended to prevent, destroy, or mitigate microorganisms or pests (e.g. pesticides).

Products in the first category are only subject to regulation under FIFRA and products in the second category are subject to both registration requirements and regulation under FIFRA. Water treatment units not intended to prevent, destroy, repel or mitigate any microorganisms or other pests are not regulated under FIFRA.

The first registration was issued in 1965 by the U.S. Department of Agriculture, the U.S. Environmental Protection Agency has been registering units since 1975. For example, there were approximately 147 water treatment products registered in 1988.[2]

For bacteriostatic water filters, the only data required is chemical data demonstrating that no more than 50 μg/L silver are released into the effluent water. These units can only be recommended for use in conjunction with municipally treated water or water that is already microbiologically potable.

Water purifiers fall in the category of pesticide products with public health-related uses because they are recommended for use on raw or untreated water or water of unknown source or quality. Bacteriological and chemical data are required for these products and are for emergency use only.

Federal Trade Commission

The Federal Trade Commission (FTC) established in 1914, is a small independent Federal agency with the mandate to promote free and open competition and protect consumers from unfair and deceptive practices. In general the FTC can act only in the broad public interest but has strong remedial powers when it finds a practice to be unfair or deceptive. Although the FTC has no specific regulations governing the advertising of water treatment devices, claims for these devices must be truthful and substantiated.[3]

Because of an increase in marketing abuses in the sale of home water purifiers, the FTC also announced that it had prepared a brochure entitled "Facts for Consumers Buying a Home Water Treatment Unit". The brochure prepared with assistance from the U.S. Environmental Protection Agency is designed to help consumers determine whether they need a water treatment unit and what various types of units can do. It alerted consumers to deceptive sales practices that unscrupulous sellers use.

An illustration of this process is the action that the Federal Trade Commission took against two Florida telemarketers for misrepresenting the capabilities of "water purifiers" they sell and the return or replacement policies for these purifiers (FTC News, 1989). Action against these two manufacturers was filed in separate complaints in federal court.

The commission also charged one of the companies with falsely telling consumers that they would receive an award without making a purchase and failing to disclose to consumers that they would have to pay substantial additional sums of money to obtain the award. A federal court in Miami issued temporary restraining orders prohibiting such misrepresentation and froze the defendants assets.

The Commission's law suits charged that the telemarketing companies sent postcards to consumers claiming that they had won an award. Consumers who responded were connected to the defendant's telemarketers who delivered a sales pitch for one of the two water purifiers.

According to the Commission's complaint, one sales pitch falsely and deceptively claimed that: the water purifier will remove or eliminate from at least 10,000 gallons of tap water 99%-100% of various impurities such as bacteria, radon gas, arsenic, asbestos or coliform. In fact, the Commission alleges that this water purifier is incapable of such performance. The Commission alleged that the water purifier is not registered with or approved by EPA as claimed by the manufacturer. The manufacturer's warranty on the water purifier provided free-of-charge replacement. In fact, according to the Commission, the consumer was forced to pay a prorated amount for replacement under warranty.

In addition, the complaint charged that the company told the consumers who purchased water purifiers that they would receive a valuable award such as $5,000.00 in gift certificates and a four day Las Vegas vacation including air fare for two without being required to make any additional purchase. However, consumers instead received vouchers that required payment of substantial additional sums of money to obtain awards. The complaint alleged that the defendants failed to disclose the required additional payments.

The Commission's complaint against the second company charged that it claimed that its water purifier would eliminate from tap water various impurities such as *Salmonella*, asbestos, pesticides, lead, mercury and PCBs. This claim was false, according to the Commission because the water purifier does not eliminate the stated contaminants.

The Commission charged that the company also claimed that consumers could cancel the purchase of the water purifier and obtain a full refund within sixty or ninety days or some other stated period of time. However, according to the FTC complaint, consumers either didn't get any refund at all or were charged a restocking fee of 10-25% of the purchase price or some other amount. The company did not disclose this "restocking fee" according to the complaint.

The Commission asked the court to issue preliminary and permanent injunctions in both cases halting the illegal behavior and also asked the court to order the defendants to pay refunds to consumers. The complaints were filed in the U.S. District Court for the Southern District of Florida - Miami Division. The Commission files a complaint when it has "reason to believe" that the law has been or is being violated and it appears to the Commission that a proceeding is in the public interest. The complaint is not a finding or ruling that the defendant has actually violated the law. The case is ultimately decided by the courts.

U.S. Environmental Protection Agency

In a November 1985 Federal Register notice, the USEPA proposed that, in general, point-of-use (POU) and point-of-entry (POE) technologies not be considered Best Technology Generally Available (BTGA) (50 FR 46916, 1985). It did allow POU/POE to be considered acceptable technology to meet MCLs if certain conditions are met. The reason EPA did not propose POU or POE technologies as BTGA was because of specific difficulties associated with these devices. A primary concern with POU/POE is the difficulty associated with monitoring compliance and assuring effective treatment performance in a manner comparable to central treatment. An additional concern is that POU devices only treat drinking water at a single tap, leading to potential exposure via ingestion at untreated taps and exposure introduced through indoor air transport, for example, from showers or dermal contact. In addition, these devices are not generally affordable by large metropolitan water systems which is one of the criteria for setting Best Available Technology (BAT).

The majority of commenters agreed that the POU/POE devices should not be considered BAT, and that the National Primary Drinking Water Regulations should not allow their use for compliance with MCLs, due to difficulties in controlling installation, maintenance, operation, repair, and potential human exposure via untreated taps. Other commenters took the position that POU/POE devices should be considered BAT or allowed for compliance, because these technologies were most often more cost effective for some small systems than central treatment.

In the final rule, POE and POU devices are not designated as BAT because: (1) it is significantly more difficult to monitor the reliability of treatment performance and to control POE and POU devices in a manner comparable to central treatment; (2) these devices are not generally affordable by large metropolitan water systems; and (3) in the case of POU devices, not all water is treated (40 CFR, 1987). POU devices are not considered acceptable means of compliance with MCLs, because they do not treat all the water in the home and could result in health risks due to exposure to untreated water. Therefore, POU devices are only considered acceptable for use as interim measures as a condition of obtaining a variance or to avoid unreasonable risks to health before full compliance can be achieved. POE devices are acceptable means of compliance because POE can provide drinking water that meets the standards throughout the home. Because these devices are often essentially the same as central treatment for small systems they may be cost effective for small systems or non-transient non-community water systems. It is recognized that operational problems may be greater than for central treatment in a community system.

Under the Safe Drinking Water Act (SDWA), EPA is required to establish necessary conditions that assure protection of public health. The Act specifically states that primary drinking water regulations are to contain "criteria and procedures to assure a supply of drinking water which dependably complies with ... maximum contaminant levels, including quality control and testing procedures to ensure compliance with such levels and to ensure proper operation and maintenance of the system. The rule therefore imposes the following conditions on those systems that use POE for compliance:

(1) Central Control. Operation and maintenance of all parts of the system is the responsibility of the public water system. As long as the public water system maintains control of the operation of the device, central ownership is not necessary. Central control is appropriate and necessary to ensure that the treatment device is kept in working order.

(2) Effective Monitoring. Each public water system must develop a monitoring plan and obtain State approval before it installs POE devices. Because POE devices present a fundamentally different situation than central treatment, a unique monitoring plan must be developed. The monitoring plan must ensure that POE devices provide health protection equivalent to central water treatment. In this case equivalent means that the water would meet all Primary and Secondary Drinking Water Standards. The water must be of acceptable quality similar to water distributed by a well-operated central treatment plant. Monitoring information, in addition to VOC data, must include physical measurements and observations, such as total flow treated and the mechanical condition of the treatment equipment.

(3) Application of Effective Technology. There are many types of POE designs available and there are no generally accepted standards for the design and construction of POE devices. The State must therefore require adequate certification of performance, and field testing. In addition, the State must require a rigorous engineering design review of each type of device. Certification can be done by the State or by a third party acceptable to the State.

(4) Maintenance of the Microbiological Safety of the Water. The tendency for increases in bacterial concentrations in water treated with activated carbon and some other technologies must be considered in the design and application of POE devices. Frequent backwashing, post-contactor disinfection, and monitoring must be practiced in order to ensure that the microbiological safety of the water is not compromised. EPA considers this condition necessary because disinfection typically is not provided after point-of-entry treatment as is normally used in a central treatment plant.

(5) Protection of All Consumers. If a building is connected to a public water system and has a POE device installed, maintained and adequately monitored, then if the building is sold, the rights and responsibilities of the utility customer must be transferred to this new owner with the title.

Department Of Housing And Urban Development

The U.S. Department of Housing and Urban Development (HUD) has issued a Mortgagee Letter (1991) to all of its regional offices which specifically prohibits the use of POU equipment. Only whole house (POE) treatment will be acceptable for a prospective homeowner to get federal financing regardless of the water contamination problem.

THE CANADIAN POSITION

Point-of-Use devices are becoming common household appliances in Canada with total sales estimated at about 100,000 units per year.[4] However, there is no specific legislation controlling the sale, use, or performance of point-of-use water treatment devices in Canada. Several regulatory acts do have provisions that could be related to these devices as described below.[4]

- The Pest Control Products Act of 1968-69. This act regulates products that are intended to control any kind of pest such as insects, fungus, bacterial organisms, viruses, weeds, rodents, or other plant or animal pests. Under this act, devices such as ozonators, chemical feeders, etc. may be covered when they are used for nonpotable water such as swimming pools, spas, etc.

- The Medical Devices Regulations (PC-1976-2031 and revised periodically) of the Food and Drugs Act. This act regulates devices that are manufactured, sold, or represented for use in the prevention of disease. If, therefore, a device claims to disinfect water and prevent enteric disease, it could fall under the provisions of this act. The regulations require the manufacturers to notify the department when a device is put on the market and to furnish certain information including a statement of purpose of the device and a copy of instructions.

- The Hazardous Products Act (1968-69) administered by Consumer and Corporate Affairs authorizes the Minister to carry out investigations and demand information regarding consumer products to determine whether such products are likely to be a danger to the health or safety of the public. There are provisions in this act to remove a product from the market.

- The Competition Act (1986) is used where misrepresentation of devices has been alleged. Section 36 of this act states that no person shall make a representation to the public that is false or misleading in a material. Also, a representation to the public in the form of a statement, warranty, or guarantee of the performance, efficacy, or length of life of a product that is not based on an adequate and proper test is forbidden.

Although these legislative tools are for more serious problems, the policy of the Department of Health and Welfare is to try to avert problems before they occur. For point-of-use devices, this program includes testing and evaluation of the devices, advice and educational materials, and cooperation with industry and non-profit organizations.

Evaluation of devices is conducted on an ongoing basis at the request of the public, governmental and non-governmental agencies, and industry. An evaluation usually involves the review of data and claims made for a device, the validity of claims, adequacy of test protocols, etc.

The Province of Ontario, like the Canadian government, does not regulate water treatment devices. However, an Ad Hoc Committee on home water treatment devices composed of members from the Ministry of the Environment, the Ministry of Consumer and Commercial Relations, Health and Welfare Canada, Ontario Research Foundation, and the Canadian Water Quality Association prepared some guidelines in 1988.[5]   The guidelines are designed to provide information on different treatment devices including conditions for use that will ensure their effectiveness and limitations on their usefulness. In addition, the guidelines suggest that claims for devices conform with the voluntary industry standards for the promotion and advertising of water treatment devices. These guidelines are informational only and not enforceable.

STATE REGULATIONS

As of November 1990, thirteen states were considering legislation that effects the POU/POE industry. The following ten states have laws in place that impact the POU industry. Some of these activities are summarized in the following sections. The California and Wisconsin programs are discussed in detail.

Because it is in the forefront in regulating POE/POU devices, the State of California rules and regulations will be described in the following section in some detail as will the actions of a number of states in regulating POU/POE devices.

California

During 1986, the California legislature introduced two legislative measures regarding point-of-use (POU) point-of-entry (POE) water treatment devices.[6] One measure, Senate Bill SB 2119 by Senator Torres, addressed the performance of POU/POE water treatment devices for which a claim relative to the health or safety of drinking water is made. The other measure, SB 2361 by Senator McCorquodale, addressed advertising claims made in the sale of POU/POE water treatment devices. Both bills were passed by the legislature, signed by the Governor, and became effective on January 1, 1987.

In general, the two bills were considered to be tough pieces of legislation. A frequently asked question is why the legislature decided to regulate the water treatment device industry. Since the California legislature does not maintain a record of their committee proceedings, one can only speculate why these bills were passed. What is known is that a number of events, preceding the introduction of the bills, had come to the attention of the legislature.

Since the late 1970's, Californians have realized that their ground water sources of drinking water were potentially vulnerable to chemical contamination. In 1978, a large number of wells in the San Joaquin Valley were found to be contaminated with the agricultural fumigant, dibromochloropropane (DBCP). Many of these wells were found to exceed the state's action level for DBCP of 1 $\mu g/L$ (1 ppb).

During 1980, wells in the heavily populated San Fernando Valley and San Gabriel Valley in Los Angeles County, were found to be contaminated by industrial solvents such as trichloroethylene (TCE) and perchloroethylene (PCE). These same industrial solvents were also detected in wells in the Santa Clara Valley, which is often referred to as Silicon Valley.

In 1985, the California Department of Health Services (DHS) sampled over 3,000 wells used by large public water systems (over 200 connections) for organic chemical contaminants. A significant number (18.3 percent) of the wells sampled had measurable concentrations of one or more organic chemicals.

One hundred and sixty five of these wells (5.6 percent) had concentrations of chemicals that exceeded the State Maximum Contaminant Level (MCL) or a State Action Level. When contamination levels were found to exceed an MCL or State Action Level, public notification was initiated by means of a public news release.

During 1985 and 1986, the newspapers and television stations, particularly in Southern California, frequently reported on drinking water contamination problems. The public was alarmed by these news stories and
became very concerned about the quality of their drinking water.

As a result of the increased public concerns about drinking water
quality, the bottled water industry recorded a significant increase in
sales in 1985 and 1986. The water treatment device industry appears to
have experienced a similar increase in sales during the same period.
Unfortunately, there were a number of cases of consumer abuse and fraud
as a result of the overly aggressive marketing efforts by a few companies
selling water treatment devices.

The introduction of SB 2361 by Senator McCorquodale has not been tied
to any specific consumer problems. However, the Senator represents the
Santa Clara Valley and his office was contacted by constituents about the
marketing techniques used by the water treatment industry in that area.

SB 2361 enacted a statue which provides "truth-in-advertising" as it
relates to water treatment devices. The statute addresses false or misleading advertising with key provisions which made it unlawful to:

- Make false claims or statements about the quality of water
  provided by a public water system.

- Make false claims about the health benefits provided by the use
  of a POU/POE water treatment device.

- Make any product performance claims unless such claims are based
  on actual, existing factual data.

- Make any other attempts to mislead the consumer or misrepresent
  the product.

This statute is expected to deter unscrupulous salespersons and reduce the number of complaints relative to fraudulent sales. The statute
will also assist the consumers who are victims of fraudulent sales by
providing them with a means to file a criminal misdemeanor action and
recover damages. The statute does not assign enforcement responsibility
to any specific agency. It is expected that local district attorneys and
the State's Attorney General will take legal action against companies
acting in violation of this law.

The introduction of SB 2119 by Senator Torres is often described as
a response to a consumer abuse problem in McFarland, California. A water
treatment device company was reported to have advertised and convinced
customers in McFarland that their water treatment device could remove all
cancer-causing chemicals. This marketing effort occurred at a time when

this community was very concerned about a cluster of childhood cancer cases that were being investigated by state and local health agencies. The company was successfully prosecuted by the State's Attorney General and the settlement allowed the consumers to rescind their sales contracts.

SB 2119 enacted a statute which requires that any water treatment device for which health benefit claim is made, cannot be sold in California unless the device had performance testing that has been certified by the Department of Health Services (DHS). This law further requires DHS to adopt regulations setting forth the criteria and procedures for certification of water treatment devices. These regulations must include appropriate testing protocols and procedures to determine the performance of these devices. The cost of this new program is to be paid for through fees imposed on the applicants. The law also assigned responsibility for enforcement to DHS or local health departments.

This statute outlines a very specific plan for the regulation of POU/POE water treatment devices (WTDs). The general provisions of this statute include the following:

- The DHS is required to adopt regulations which set forth the criteria and procedures for the certification of WTDs that are claimed to affect the health and safety of drinking water.

- A "water treatment device" (WTD) is defined to mean any point-of-use or point-of-entry instrument or contrivance sold or offered for rental or lease for residential, or institutional use, without being connected to the plumbing of a water supply intended for human consumption in order to improve the water supply by any means, including, but not limited to, filtration, distillation, adsorption, ion exchange, reverse osmosis, or other treatment.

- No WTD which makes product performance claims or product benefit claims that the device affects health or the safety of drinking water, shall be sold or otherwise distributed unless the device has been certified.

- WTDs which are not offered for sale or distribution based on claims of improvement in the healthfulness of drinking water need not be certified.

- A WTD initially installed prior to the operative date of the statute is not required to be certified.

- The requirement that a WTD be certified does not become operative until one year after the effective date of the regulations.

- The DHS or any testing organization designated by the DHS may
  agree to evaluate test data for a WTD offered by the manufacturer, in lieu of the requirements of the statute, if the DHS
  or the testing organization determines that the testing
  procedures and standards used to develop the data are adequate
  to meet the requirements of the statute.

- The DHS may accept a WTD certification issued by an agency of
  another state, by an independent testing organization, or by
  the Federal government in lieu of its own if the DHS determines
  that the certification program meets the requirements of the
  statute.

The provisions that are to be included in the DHS regulations were
defined in the statute with considerable detail. The provisions that are
required or allowed as part of the regulations are as follows:

- The regulations shall include appropriate testing protocols and
  procedures to determine the performances of WTDs in reducing
  specific contaminants from public or private water supplies.

- The regulations may adopt, by reference, the testing procedures
  and standards of one or more independent testing organizations
  if the DHS determines that they are adequate to meet the requirements of the statute.

- The regulations may specify any testing organization that the
  DHS has designated to conduct the testing of WTDs.

- The regulations are required to include minimum standards for
  (a) performance requirements, (b) types of tests to be performed, (c) types of allowable material, and (d) design and
  construction.

- The regulations are required to include requirements relative
  to product instructions and information, including product
  operation, maintenance, replacement, and the estimated cost of
  these items.

- The regulations may include any additional requirements, not
  inconsistent with the statute, as may be necessary to carry out
  the intent of the statute.

Finally, the statute specifies procedures for the enforcement of the
act. Key enforcement provisions include the following:

- The DHS, or any local health officer with the concurrence of
  the DHS, is responsible for the enforcement of the act.

- The DHS may suspend, revoke, or deny a certificate upon its determination that either (a) the WTD does not perform in accordance with the claims for which certification is based, or (b) the manufacturer, or any employee or agent thereof, has violated the statute.

- The act provides that any person, corporation, firm, partnership, joint stock company, or any other association that violates any provision of the act, is liable for a civil penalty not to exceed $5,000 for each violation.

The DHS Public Water Supply Branch (PWSB) has been given the responsibility for the implementation of SB 2119 and is currently developing policy and regulations for the implementation of the certification program. The PWSB has established an informal advisory committee consisting of representatives from industry, water utilities, and consumers to assist in identifying and addressing issues. The following are some of the elements of the program and regulations that are being considered:

- The DHS plans to adopt existing protocols and standards such as those established by the National Sanitation Foundation (NSF).

- A "health or safety claim" will be defined in terms of the Primary Drinking Water Standards adopted by the DHS or the U.S. Environmental Protection Agency.

- Certification of a WTD will be based on specific contaminants for which the manufacturer has made a health or safety claim.

- The DHS will not establish a state laboratory to conduct the testing required for certification. The DHS plans to contract with outside laboratories or testing organizations for the testing and other administrative tasks relative to certification.

The water treatment device industry is very concerned as to how the WTD certification program will impact the marketing and sales of their product in California. The advisory committee has been very helpful in bringing the industries concerns to the attention of the PWSB. Some of the concerns that have been identified are as follows:

- The failure to accept manufacturer's data would impose a substantial cost on the industry.

- If retesting by a third-party laboratory or testing organization is required, the manufacturers will have to pass on the added expense to the consumers.

- The cost of testing under NSF or equivalent standards will be very expensive.

- The one year grace period in which all testing must be completed may exceed the capacity of State contract laboratory or laboratories designated to conduct WTD performance testing.

- In order to reduce the costs associated with performance testing, consideration must be given to testing approaches such as the use of surrogates and the extrapolation of data whenever possible.

It is evident that the California legislature has given the DHS a difficult assignment. However, the Department is committed to the establishment of a WTD certification program that will serve the needs of the California consumers and still be responsive to some of the unique problems of the water treatment device industry. The DHS is also confident that the California program will not be in conflict with any efforts to establish a national certification program.

As indicated previously, a driving force for regulation of POU/POE devices is the interest in these devices by the general public. Part of the reason for this interest in California is that studies conducted by the Department of Health Services have shown wide-spread chemical contamination of ground water. Publicity over the presence of potentially cancer causing compounds in drinking water has fueled consumer fears. As a result of this publicity and concern, manufacturers and distributors of POU/POE devices have begun to step up their sales campaigns. Visits by unscrupulous salesman have caused citizens to complain about the use of misleading advertisements. A consequence of these types of problems has been the loss of creditability of public water systems.

The general position of the Department of Health Services regarding point-of-use is outlined below:

1.   Unless the feasibility of all other available alternatives has been properly evaluated, the Department will not recommend the installation of point-of-use devices by community water supply systems. A comprehensive report which contains the following items must be prepared by the utility and submitted to the Department for review.

   a.   The water quality problems experienced by the utility and the attempts that have been made to correct them must be evaluated.

   b.   The possibility of interconnecting with an adjacent utility which has potable water must be evaluated.

    c.   If point-of-use treatment of surface water supplies is proposed, the utility must investigate the development of groundwater sources.

    d.   The utility must research the availability and cost of properly treating surface water sources before point-of-use devices are considered for a contaminated groundwater supply.

    e.   The possibility of deepening an existing well or constructing properly sealed wells to avoid contaminants must be evaluated.

    f.   The cost and problems associated with installation and operation of point-of-use versus providing central treatment of the entire supply must be evaluated.

    g.   A time schedule and a comprehensive economic evaluation must be made for implementation for each of the above-listed alternatives.

2.   Only conditional approval by the Department will be granted for point-of-use treatment. In the event new treatment techniques are developed which are more economical, the utility would then be required to provide central treatment of the sources.

Listed below are several examples of conditions which should be considered in the approval process for the point-of-use devices.

1.   The utility must be advised that they are still responsible for monitoring all sources of supply as presently required by State and Federal law.

2.   Ongoing monitoring of the contaminants must be performed to verify that the concentrations have not changed appreciably. This information will be used to evaluate the source water quality and the ability of the treatment units to operate properly.

3.   The utility is responsible for monitoring and maintaining all point-of-use devices installed within the system.

4.   Point-of-entry devices must be utilized for organic compounds because inhalation and skin absorption are possible exposure routes.

5.   Individual treatment units will be owned by the utility and access to point-of-use devices for monitoring and maintenance purposes must be provided.

6. As required by SB 2119, the device selected will have to be certified by the Department of Health Services.

7. Pilot plant studies must be performed which document that the device selected is able to continuously remove the contaminant in question for the needed length of time. This will be covered by certification testing.

8. Adequate contact time must be provided for disinfection when granular activated carbon units are utilized for organics removal or when cartridge filters are used for surface water treatment.

9. The minimum system components for each installation may include a totalizing water meter that also provides instantaneous flow, cartridge prefilter, pressure gauges, sampling taps and gate valves.

10. The utility must collect treated water samples routinely for analyses to verify proper operation of the devices. In addition, raw water samples and treated water samples for bacteriological quality must be collected at least monthly from a representative number of sites. The utility is responsible for all system monitoring.

11. Monthly reports must be submitted which document the performance and maintenance provided.

12. Although media replacement is site specific and depends on the volume of water treatment, the minimum frequency of replacement must be established prior to device approval.

13. The utility will be responsible for the disposal of the replaced unit.

14. The utility will be responsible for development and implementation of an operation's manual which outlines the procedures to be used by the utility in the operation and maintenance of these devices.

15. The utility should investigate similar installations to obtain information with regard to problems that have been experienced. For instance, if another state has a utility that employs point-of-use devices for removal of the contaminant, an investigation into the success or lack of it by this utility should be made prior to giving approval for a similar installation in the state of California.

Since water within the distribution system will not be potable, the question of liability must be considered. Consumption of untreated water could occur and may result in either illness or lawsuits, therefore, the liability of the utility to new homeowners must be investigated. The utility must routinely notify their customers of the impacts associated with drinking the raw water and of the maintenance which must be performed on their home treatment unit.

The State of California Department of Health Services has been revising its proposed regulations that address product testing/ registration.[6] These regulations will probably go into effect early in 1991, and include five areas:

- Mandated product testing for health-reacted contaminants by an independent laboratory

- Initial and annual certification

- Five-year product retesting and certification

- Mandated performance data sheet

- Filing fees (per model)

  $1400 initial certification or recertification
  $400 annual renewal
  $200 late renewal penalty
  $300 modification or certification

The California legislature has passed Senate Bill 2334, which is intended to regulate the efficiency of reverse osmosis systems. According to the bill after January 1, 1991, no RO system shall be sold, installed or rented for residential use unless it is equipped with a shutoff valve or, through other equipment design specifications, achieves or exceeds equal or greater water savings than would occur with an automatic shutoff device.

All RO systems sold, installed, rented or under service contract prior to the January 1, 1991 date, must be retrofitted with an automatic shutoff or an equivalent equipment design.

Connecticut

A law was passed that prohibits brine discharge from home treatment units into private septic systems.[1]   Also, Bill No. 838 which was referred to the Committee on Public Health proposed creation of a task force to study the sale and use of certain devices for potable water.

Florida

Senate Bill 1100, requires mandatory water testing of private well water but adjustment to water quality however, is left to the discretion

312 POU/POE FOR DRINKING WATER TREATMENT

of the owner.[1]   This bill has been passed and sent to the Governor. Legislation has passed which outlines misleading advertising in the sale of water treatment devices. Information must be provided to the consumer which indicates that plans for operation, maintenance, and replacement are required for the devices to perform satisfactorily.

Iowa

Provisions under the Iowa Residential Water Treatment Law dealing with POU/POE technology are as follows:[1]

- Mandated product testing for health contaminants by a third party or manufacturer

- Temporary, initial and annual registration

- Consumer information pamphlet

- Performance data sheet

- Fees

    $400 per model initial registration
    $200 per model annual registration, which must be filed between
        April 1 and June 30 of each year.
    $200 per protocol for review and approval

Massachusetts

The state's plumbing board has begun to enforce an old law requiring the testing of equipment.[1]

New Jersey

The state of New Jersey has a law prohibiting brine discharge into septic systems.[1]

NEW YORK

Assembly Bill 3765 prohibits false and misleading advertising regarding water treatment products or a consumer's drinking water supply.[1] It also requires that a performance data sheet be provided to the purchaser, renter, or lessee prior to the consummation of any sale of a water treatment unit.

In the case of a catalog sale, the consummation of a sale occurs three days after the purchaser has received the water treatment unit. As of January 1, 1991, AB 3765 requires that all products be tested by a qualified laboratory.

Oregon

The State's Plumbers Board has initiated a Joint Apprentice Training Council (JATC) license for water treatment companies.[1] The requirements of the sub-license will apply to those who install water treatment equipment.

Tennessee

In 1988, Tennessee became one of the first states to enact legislation addressing misleading advertising.

Wisconsin

In June 1988, Chapter ILHR 84, of Wisconsin's Plumbing Code was changed to reflect a major reorganization of the statewide plumbing product review panel.[1]

The major provisions relating to water treatment devices include:

• Product testing for health-related and aesthetic contaminants by an independent testing laboratory or the manufacturer

• Product review by the department of Industry, Labor and Human Relations

• Product registration

• Fees

  $100 per unit (initial)
  $50 per unit (renewal)

Wisconsin's involvement in regulating point-of-use and point-of-entry water treatment devices involves five different state agencies.[7] Two of those state agencies, the Department of Justice (DOJ) and the Department of Agriculture, Trade and Consumer Protection (DATCP), have consumer protection sections. Water treatment device manufacturers and dealers only become involved with DOJ and DATCP if their advertising literature or sales practices appear to be false or misleading.

A third state agency, the Department of Health and Social Services (DH&SS), is responsible for recommending enforcement standards for ground water contaminants of public health concern. The enforcement standard may be the actual maximum contaminant level (MCL) set by the United States Environmental Protection Agency or may be below the MCL. When an MCL has not been set for a contaminant of public health concern, the enforcement standard will establish the upper limit concentration for the contaminant in ground water.

Water treatment device manufacturers and dealers rarely become involved with any DH&SS activities. However, the Department of Natural Resources (DNR), the fourth state agency, uses the enforcement standards to establish whether or not a water supply is contaminated. After a public hearing process, the DNR usually adopts the recommended enforcement standard into its regulations. If a water supply contains a contaminant of public health concern in excess of an enforcement standard, the water supply is deemed contaminated. The DNR develops regulations for methods to be pursued in obtaining pure or non-contaminated drinking water for human consumption.

If a water supply contains a contaminant in excess of an enforcement standard, the DNR requires the owner of that water supply to first seek a naturally safe water supply which can involve:

- Extending a well casing
- Drilling a new well, or
- Connecting to a public water supply or other non-contaminated well

The DNR requires all water for human consumption to be non-contaminated. Department of Industry, Labor and Human Relations regulations essentially require all water going to plumbing fixtures to be non-contaminated. Point-of-use or point-of-entry water treatment devices used to reduce the concentration of contaminants below the enforcement standard may only be installed upon approval of the DNR. The DNR also has the authority to require sampling and maintenance for these water treatment devices. Point-of-use devices are usually not designed to produce the volume or flow rate of non-contaminated water needed and so at this time are not allowed for use on contaminated water supplies. The DNR also considers point-of-entry water treatment devices at this time to be the last resort or at best an interim solution until a naturally safe water supply can be obtained.

Water treatment device manufacturers and dealers may become involved with DNR regulations if they want their devices to be used to reduce the concentration of a contaminant below the enforcement standard.

The fifth state agency is the Department of Industry, Labor and Human relations (DILHR), which reviews all point-of-use and point-of-entry water treatment devices for the following:

- Rendering inactive or removing aesthetic and health related contaminants

- Suitability of construction materials for use with potable water

• Ability of the device to withstand the pressures to which it will be subjected, and

• Proper installation instructions

The Wisconsin Department of Natural Resources (DNR) has developed a POE treatment plan approval process specifically for the removal of radium. The issues dealt with are legal and economic liabilities, effective technology, operation and maintenance, monitoring, and department approval. Under these guidelines, the water system owners are held responsible for the quality provided at each consumer's tap. Any point of entry treatment device approved must provide treatment and monitoring protection equivalent to centralized water treatment.

The criteria that is considered in a plan proposed to the DNR for use of point-of-entry treatment are as follows:

1. Legal and Economic Liabilities

    Water system owners are responsible for the quality of the water provided at each customer's tap. Any point-of-entry treatment plan approved must provide treatment and monitoring protection equivalent to centralized water treatment. Specific owner legal and economic liabilities associated with POE treatment include:

    a. System owners are responsible for owning, installing, servicing, monitoring, replacing or updating as necessary, all treatment units. Responsible third parties may be hired at the owner's discretion and with DNR approval to execute some or all of the installation, operation, maintenance, and monitoring responsibilities. It is also the system owner's responsibility to assure proper disposal of the used media. System owners will have to purchase/obtain ownership of any existing privately owned treatment units intended to be used for radium removal.

    b. Treatment units must be installed to treat all incoming water (point-of-entry) to assure protection at every tap.

    c. Responsibility for maintaining the microbiological safety of the water based on the treatment device(s) selected.

    d. Responsibility to assure that the rights and responsibilities of the customer transfer upon sale of property to a new owner.

e.   Responsibility to assure that all treatment solids and/or liquid wastes created are properly disposed of in accordance with the Department's "Interim Guidelines for the Disposal of Liquid and Solid Waste Containing Radium from Wisconsin Water Treatment Plants", or any subsequent modifications or additions to the guidelines.

f.   Similar responsibilities, as above, for any necessary associated treatment devices such as iron removal equipment.

g.   It is recommended that system owners investigate, with their legal counsel, their legal liability should POE treatment be installed.

2.   Effective Technology

The following discussion is presented assuming ion exchange softening devices will be used, however, other applicable treatment technologies may also be considered.

It will be necessary to provide verification to the DNR that the treatment unit(s) proposed, including any in-line hardness testing devices, will provide adequate radium removal while not adversely effecting other water quality parameters. Water supply owners will not be allowed to install softening methods for radium removal in buildings where untreated corrosive water will result. Remedial action to control a corrosive water situation resulting from softening is most appropriately handled through centralized treatment by the system owners.

Pilot plant/field testing will be required by DNR prior to final plan approval. Similar requirements are necessary for all existing POE treatment units to be continued in service for radium removal purposes. All treatment units must be approved by the Department of Industry, Labor and Human Relations.

Upon start-up of POE treatment facilities, all treatment units and installations must be inspected to verify the following:

a.   That every applicable premise has been provided with the required treatment facilities.

b.   That the treatment unit(s) have been properly installed in the building plumbing and that all drinking water taps are covered by treatment.

c.   That the treatment unit(s) and any associated bypass water regulation device, water hardness detection device, etc. are functioning properly.

d.   That the finished water is tested for radium content. Testing for a surrogate radium parameter such as hardness may be allowed as approved by the Department.

e.   That the finished water is tested for other water quality parameters as necessitated by the type of water treatment installed. Examples might be finished water corrosivity and iron content. Collection of a safe bacteriological sample will be required.

3.   Operation and Maintenance

Operation and maintenance service must be provided by the owner for each unit installed. Large numbers of units and limited access to privately owned dwellings are expected to be problems. Some evening and weekend inspections of treatment installations will likely be necessary. The system owner's legal liability would be expected to increase when entering and working in privately owned buildings.

POE units can be tampered with or neglected by users, leading to ineffective treatment. System owners will be required to take all necessary actions to maintain effective treatment at consumer premises where owners object to treatment installation or where tampering with treatment units is encountered.

4.   Monitoring

The water distribution system must be monitored for radium content on an annual basis.

a.   An inspection of the treatment unit(s), including any associated plumbing devices and the building plumbing to assure compliance with the originally approved installation.

b.   Add sufficient salt to the brine tank for the upcoming month. Services with higher water demands may necessitate more frequent inspections and salt addition.

c.   Sample the premise for finished water radium content. Surrogate parameters (such as hardness) to approximate radium concentration may be allowable if demonstrated effective through pilot plant or field testing.

Where surrogate monitoring is approved, all units shall be monitored for finished water radium content at least once every four years. Premises sampled for radium must represent a composite of treatment units of varying age and manufacturers unless similar units were purchased and installed upon start up of treatment. Different premises must be sampled each year for radium so that eventually all premises are sampled. The sampling and analytical costs must be borne by the system owner.

The water system owner will be responsible for providing a yearly report to the appropriate District office of the Department. The report must provide the summary of annual and monthly inspection and sampling results. A report format will be required to be submitted for approval along with the treatment plant and specifications.

5.   The Department will not approve a point-of-entry treatment proposal for radium removal unless POE treatment has been compared to other possible radium reduction alternatives in an engineering report. The other alternatives would be alternate water sources, blending of water sources and centralized treatment alternatives. Included in the engineering review must be comparisons of radium removal efficiencies, treatment reliability, waste disposal concerns, costs, operation and maintenance concerns, and general feasibility.

DNR will only take under consideration any engineering proposal for radium removal by point-of-entry treatment which completely addresses all of the above concerns. The proposed treatment program must provide radium removal protection equivalent to centralized treatment and the compliance monitoring required under those circumstances. Demonstration of a workable program will be required. Each proposal will be reviewed on a case-by-case basis. The Department's requirements for point-of-entry treatment will be revised and updated as necessary to reflect practical experience and any new State or Federal requirements.

In applying these regulations, the State of Wisconsin obtained a restitution settlement in 1990 from National Safety Association (NSA) of Memphis, Tennessee for consumers who bought its products. In the consent judgement, NSA paid $30,000 in civil forfeitures, offered to buy back unsold units in its dealers' inventories, and agreed to cease the following practices:

- Misrepresenting the quality of a customer's water
- Making a performance claim for which the product has not been approved
- Making claims of potential earnings for dealers without disclosing the basis for the claim
- Misrepresenting earnings potential by not disclosing what the actual earnings experience has been
- Selling unapproved products
- Misrepresenting the need for a plumbing permit

Iowa Consumer Guide

In 1988, the Iowa legislature passed the Iowa Consumer Protection Law [Iowa Code Chapter 714 and Iowa Administrative Code 641-14.1 (174)] which was intended to ensure that sellers and manufacturers of residential water treatment systems (POU/POE) made reasonable claims about the performance of their systems in removing contaminants from drinking water. The three major requirements of the law were: (1) the manufacturers or sellers of treatment systems which claim to remove health related contaminants must test their units according to state approved procedures; (2) sellers of treatment systems must provide these test results on a performance data sheet for each unit; and (3) the Iowa Department of Public Health is required to develop a consumer information pamphlet to help consumers make an informed purchase if they wish to buy a residential water treatment system.

The appropriate performance data sheet and the consumer information pamphlet must be given to buyers prior to their purchase of a water treat-ment unit. The seller and the buyer must sign the performance data sheet and the seller must retain a copy of the sheet as proof that it has been given to the buyer. The Iowa law applies only to residential water treat-ment devices that claim to remove health related contaminants and does not guarantee the performance of a particular treatment unit or system. It directs that the sellers must provide specific information to buyers so that they can determine whether a given system fits their situation and will work as claimed.

The consumer pamphlet describes the contaminants that may get in the water and the risks that are associated with drinking water contaminants. It then describes how contaminants can be kept out of water and why it needs to be treated. Finally, it discusses the effectiveness of various kinds of water treatment units. The pamphlet describes physical filters, activated carbon units, reverse osmosis units, distillation units, ultraviolet and disinfection units (chlorinators, ozonators), and then describes the procedures for selecting a residential water treatment system.

According to the consumer guide, the following guidelines should be used to aid the selection process:

1. Narrow the likely possibilities by reviewing the types of treatment equipment on the market and their general potential to solve a particular water quality problem.

2. Review the performance capabilities and the warranty information of various brands or models using the manufacturer's performance data sheet.

3. Select a device with the performance capabilities and warranty that matches the needs, remembering the following items:

       a.   No one treatment system corrects all water quality problems.

       b.   All systems have limitations and life expectancies.

       c.   All systems require routine maintenance and/or monitoring.

       d.   Match the treatment system to the specific contaminant to be removed.

4.   Properly install and maintain the water treatment devices.

5.   Monitor the equipment performance with periodic water testing to see if the unit is operating effectively.

The consumer guide also provides some guidance in how to interpret the performance data sheet. Information required is listed below:

1.   The name, address, and telephone number of the seller.

2.   The name, brand, or trademark of the system and its model number.

3.   The installation instructions, procedures and requirements necessary for proper operation of the system, and the seller's warranty limitations must be on the performance data sheet, or the seller may list a reference to the owner's manual for this information.

4.   Performance and test data showing the list of contaminants that the treatment system has been shown to reduce, the EPA MCL for this contaminant, the percentage reduction or test ending concentration of each contaminant, the capacity of the unit in gallons or in the period of time during which the system is effective, and if applicable, the flow rate, pressure, and temperature of the water during the testing. (The test starting concentration is the amount of the contaminant found in the water used to perform the test. The test ending concentration is the amount of contaminant found in the water after the test).

5.   The performance data sheet may list performance and test data for non health-related substances, but they may not be referred to as contaminants or health-related contaminants. Many data sheets will list these substances under "Aesthetic Effects".

The following is an example of what may be found on a performance data sheet:

REDUCTION OF HEALTH RELATED CONTAMINANTS

Contaminant	EPA Maximum Contaminant Level	Test Starting Concentration	Test Ending Concentration	Percentage Reduction
Lead	0.015 mg/liter	0.15 mg/liter	0.015 mg/liter	90%

The EPA MCL for the contaminant, the test starting concentration, and the test ending concentration or percentage reduction of the contaminant should be examined very carefully because they are part of the key to determining the effectiveness of the water treatment unit. The test starting concentration should be compared to the level of the particular contaminant in the individual consumers water. If the amount of the contaminant in a given water is much higher or lower than the amount of the contaminant in the test water, then it cannot necessarily be assumed that the treatment unit will reduce the level of the contaminant of a given water by the same percentage or by the same amount. The test ending concentration should be evaluated to determine if the unit reduced the amount of contaminant to the EPA MCL. If the data sheet lists only the percentage reduction and not the test ending concentration, then the test ending concentration can be calculated with the following formula:

Test Starting Concentration X Percentage Reduction =
    Test Concentration Removed

Test Starting Concentration - Test Concentration Removed =
    Test Ending Concentration

The information on the performance data sheet should be used to determine if the unit would be operated under the same conditions of water pressure and temperature or pressure. If this is considerably different from those in the test, the same results should not be expected. The water pressure, for example, is particularly critical to the performance of a reverse osmosis unit. In addition, the capacity of drinking water usage in gallons per day should be computed to determine how long this particular unit would be effective in the home. If it would only be effective for a short period of time, it may not be the right unit.

The seller of the unit should be asked about other conditions in a given water that may interfere with the performance of the water treatment unit. For example, extremely hard water or a large amount of iron in water can interfere with some water treatment processes.

The performance data sheet should be used to be certain that the water treatment system being considered would actually meet a given water treatment need. The performance data sheet must be given to the buyer and must be signed by the seller and the buyer before the sale is finalized.

Indiana

The Indiana Department of Environmental Management, Office of Environmental Response has provided a work sheet which gives the minimum specifications required for the design and maintenance of carbon filtration systems for removal of low levels of volatile organic compounds, refined petroleum products, and common industrial solvents from potable water supplies. The standards were developed by reviewing current systems in use and their effectiveness against the above mentioned pollutants. Also technical data from various contractors and the Indiana plumbing code was used to develop the parameters discussed. The items covered minimal mechanical specifications for POU/POE systems. These systems have been provided to home owners who where effected by a ground water contamination incident in Elkhart, Indiana.

The system was installed with a mid-point tap to enable a sample to be periodically drawn by a representative of the Indiana Department of Environmental Management (IDEM). This sample will be analyzed to determine when the homeowner should replace a carbon tank. The IDEM is currently setting up a program to provide the homeowner with limited assistance in paying for maintenance costs. In order to have this system operational as soon as possible, homeowners were requested to fill out and sign a form within two weeks after receiving a letter.

Colorado

Colorado had developed a cease and desist order for companies that market POU/POE devices within the state of Colorado. By February 1989, the Attorney General for the state of Colorado had obtained 29 cease and desist orders preventing various deceptive advertising and sales practices by local and out-of-state individuals and companies. The cease and desist orders addressed the scare tactics used by marketers of water purification systems. In May 1988, the Attorney General and the Colorado Department of Health Drinking Water Section launched an investigation after receiving complaints from the EPA, numerous concerned citizens and several front ranged districts including Denver, West Minster, Brumfield, Loveland, and Ft. Collins. The complaints expressed concern over fliers, newspaper advertisements and brochures that claimed that various POU devices would filter and purify water as it comes from the tap. Many of the companies used scare tactics in their attempt to market and sell the devices. The Attorney General required companies to provide reliable scientific data to substantiate their advertising claims that drinking water in the state of Colorado was hazardous.

To this date, none of the individuals or companies responsible for the advertising has produced adequate substantiation or evidence that Colorado drinking water is contaminated.

Connecticut

The State of Connecticut has been evaluating the use of granular activated carbon (GAC) filters as an effective means of removing EDB from drinking water. After discussions with officials at EPA, the State of Florida, the University of Connecticut and water treatment companies, the State is recommending the use of GAC for the removal of EDB. They are recommending a filter containing a minimum of one cubic foot of GAC media to be used, although there may be instances where larger units or a second GAC filter will be necessary. It should also be noted that this installation provides for the treatment of all water used in the home. To make this treatment effective, Connecticut requires that virgin GAC media be used in all installations; the use of reconditioned GAC media is not acceptable. A typical one cubic foot unit measures 10 inches in diameter and four feet in height.

The Department of Environmental Protection (DEP) has issued orders to parties believed to be responsible for EDB contamination. These orders require that the polluter provide affected residences with bottled water as a short term solution and also with a new source of water supply or a form of water treatment (including maintenance and monitoring) as a long term solution. The water systems and the homeowner should realize, however, that if the homeowner wishes to install a GAC unit prior to an order being implemented, they are doing so at their own expense, since the orderee is under no legal obligation to reimburse affected parties for corrective action those parties have taken by themselves. Should the homeowner wish to proceed, the State is recommending that the town conduct a sanitary survey at each well for which GAC treatment is to be installed. As part of the survey the State asks that bacteriological, physical and chemical water samples be collected and analyzed. A review of that water quality will be useful in evaluating the need for a prefilter or disinfection.

With respect to testing the GAC units for their effectiveness in removing EDB, the homeowner should be informed that periodic monitoring of the filter is necessary and that it is the homeowner's responsibility to have tests done at a private laboratory. By reading the flow meter over a period of time, the homeowner should have a good feel for when the media needs to be replaced. With respect to the disposal of exhausted GAC media, the State prefers to leave that responsibility with the filter installer. DEP will arrive at a policy relative to the disposal of GAC media containing EDB.

REFERENCES

1. Theisen, P. M., "Proliferating POU Regulations", *Water Technology*, 13(11):26-30, (November 1990).

2. Douglas, R., "Regulatory Requirements For Point-of-Use Systems", in Proceedings from the Conference on Point-of Use Treatment of Drinking Water, U.S. Environmental Protection Agency, Water Engineering Research Laboratory, Cincinnati, Ohio 45268, EPA/600/9-88/012, pp 10-11, June, 1988.

3. Winston, J., "Federal Trade Commission Regulation of Water Treatment Devices", in Proceedings from the Conference on Point-of Use Treatment of Drinking Water, U.S. Environmental Protection Agency, Water Engineering Research Laboratory, Cincinnati, Ohio 45268, EPA/600/9-88/012, pp. 25-26, June, 1988.

4. Tobin, R. S., "Control of Point-of-Use Water Treatment Devices in Canada: Legal and Practical Considerations", in Proceedings from the Conference on Point-of-Use Treatment of Drinking Water, U.S. Environmental Protection Agency, Water Engineering Research Laboratory, Cincinnati, Ohio 45268, EPA/600/9-88/012, pp 12-14, June, 1988.

5. "The Quality of Drinking Water in Toronto", Technical Report, A Review of: Tap Water, Bottled Water, and Water Treated by a Point-of-Use Device, (September 1990).

6. Burns, R. F., "The Regulation of Water Treatment Devices in California", in Proceedings from the Conference on Point-of Use Treatment of Drinking Water, U.S. Environmental Protection Agency, Water Engineering Research Laboratory, Cincinnati, Ohio 45268, EPA/600/9-88/012, pp 15-17, June, 1988.

7. Trapp, L., "Wisconsin Regulation of Point-of-Use and Point-of-Entry Water Treatment Devices", in Proceeding from the Conference on Point-of Use Treatment of Drinking Water, U.S. Environmental Protection Agency, Water Engineering Research Laboratory, Cincinnati, Ohio 45268, EPA/600/9-88/012,pp 18-21, June, 1988.

CHAPTER 9

CONSUMER PROTECTION: WHAT STEPS TO TAKE?

INTRODUCTION

Home water treatment units are purchased for many different reasons. Some consumers may be concerned about chemicals or particulate matter that affect the taste or appearance of their drinking water. Other consumers may be concerned that their water contains harmful organisms, such as bacteria, or chemical pollutants, such as pesticides or industrial solvents. Whatever the reason for buying a home water treatment unit, consumers should consider two points: (1) if local public drinking water supplies meet national/state standards, home treatment is seldom needed for health protection and (2) no single unit will solve all varieties of water problems.

There are several questions the consumer needs to have answered before purchasing a water treatment unit.[1]

- How do I know if I need a unit?

- What kinds of treatment units are available to address my home water problem?

- What are some selection options?

- How can I protect myself from deceptive sales practices?

WHAT TO TEST FOR?

Although many home water treatment dealers are legitimate, some are unscrupulous (Figure 1). Some dealers may falsely claim that the drinking water contains harmful levels of chemical contaminants, such as chlorine or lead. Although certain communities may experience various levels of water contamination, other communities may not. Therefore, anyone who suspects that they have a water quality problem should have their water analyzed by their local health department or water company (if connected to a public supply).

Figure 1.   Unscrupulous Point-of-Use Salesman

Under the Safe Drinking Water Act, all public water supplies are required to meet the drinking water standards established by the USEPA. More stringent standards may be established by state governments. For information about the quality of drinking water, contact your local water utility offices and ask for results from the distribution system as well as the finished water as it leaves the treatment plant, especially if lead or bacterial contamination is a concern.

In contrast, private well owners are subject only to state and local laws, and they are responsible for the quality of water from their wells. For assistance with possible drinking water problems in a private well, contact your local health department. In addition, your drinking water may need to be analyzed by a state-approved testing laboratory.

Well water should be tested periodically for bacteria, inorganic compounds, and radon. Organic chemical testing should be conducted if the well is within a mile or two of a gasoline station or refinery, a chemical plant, a landfill, or a military base. In agricultural areas, have the water tested for nitrate and pesticides. One should also test for lead if their house is <10 or >30 years old or if the plumbing pipes are joined with lead solder.

Bacteriological tests consist of a coliform organism count for 100 milliliters (ml) of water while the nitrogen series consists of water analyses for the presence of ammonia nitrogen and nitrate nitrogen expressed as milligrams of constituent per liter of water (mg/L). Presence of a detergent in a water supply is an indicator of possible contamination from sewage. Sometimes the expression "parts per million" (ppm) is used to describe a water analysis. Parts per million (ppm) is the same as milligrams per liter, i.e., 1 ppm = 1 mg/L.

The following testing guide can be used if there are nearby sources of possible contamination which may be of concern (Table 1):

TABLE 1. RECOMMENDED TESTING

Possible Contaminating Source	Water Test Suggested
Sewage disposal system	Bacteriological, nitrogen series, detergents
Agricultural operations	Nitrates, pesticide scan (The scan is made only for the pesticides being used for the agricultural practice. If pesticides are being used around the home, include these in the scan.)
Sanitary landfill	Total organic carbon (TOC), total organic halogens (TOX), conductivity, pH, total dissolved solids (TDS), specific organics
Industrial wastes	Synthetic organic chemicals (SOC), specific organics
Metals, e.g. Lead	Request specific test
Radon	Radon

For new house construction or when buying a home that is on a private water well, a test of potability is recommended. The following water determinations should be conducted: coliform, iron, manganese, chloride, hardness, alkalinity, ammonia nitrogen, nitrate, pH, color, odor, and turbidity. This test series is used for the certificate of occupancy requirement and in some instances is necessary to satisfy lending agency approval requirements. If at this time there are other water health concerns, for example a nearby abandoned dump site, one should include additional water tests (e.g., SOC scan) that help ensure a safe drinking water supply. You may want to make a land use survey to help determine the types of tests that need to be conducted.

There are no set recommendations on how often to test the water. Local conditions, the type of water well (shallow or deep), where the well's water is coming from, and health concerns are all determining factors. If water test information is available on the physical and chemical characteristics, then a bacteriological test every three years may be adequate. With a shallow water well that is not adequately constructed or if the well is in a location subject to possible contamination, an annual water test is suggested. If there appears to be a noticeable change in the drinking water quality for some unknown reason, water analyses may be needed to determine the cause or if in fact there is a problem.[2]

Water analyses for health related water problems should be performed by a state approved water testing laboratory. Information on the nearest state approved laboratory can be obtained from the local health department or local Cooperative Extension System office. Aesthetic or nuisance tests can be performed by or through water equipment/conditioning dealers, major retail stores selling such equipment, or state approved water testing laboratories. A report by Consumer Reports offers the following advice:[3]

- Companies that sell water-treatment equipment often offer a free or low-cost water analysis as part of the sales effort. Don't depend on that kind of test: It's like asking a barber if you need a haircut. Consult an independent, state-certified lab instead. You can often find one in the Yellow Pages under "Laboratories - Testing".

- Or use a mail-order lab. The labs send you a kit containing collection bottles and detailed instructions. You collect water samples and ship them back by overnight package delivery. The labs provide test results and an explanation of the numbers two to three weeks later. Costs for testing range from $20-30 for a lead test, $75-100 for a combination of minerals and bacteria, to well over $100 for VOC or pesticide scans.

- The mail-order lab reports may be a little technical but not too hard to understand. State or local health officials can help interpret water analysis results.

- No single water test is perfect. Over the years, Consumer Reports has found that all labs tend to overstate or understate results occasionally.

- If a test report says your water has an especially high level of a contaminant like lead, nitrate, or radon, have the water tested by a second lab before taking costly remedial action.

The following are mail-order laboratories that have been evaluated by Consumer Reports:

- WaterTest
  33 South Commercial St.
  Manchester, NH 03101
  (800)426-8378

-    National Testing Laboratories
     6151 Wilson Mills Road
     Cleveland, OH 44143
     (800)458-3330

-    Suburban Water Testing Laboratories
     4600 Kutztown Road
     Temple, PA 19560
     (800)433-6595

WHAT TREATMENT UNITS AND OPTIONS ARE AVAILABLE?

Detailed discussion of available treatment units has already been presented in Chapter two. The following questions are those that the consumer should ask the water treatment vendor once the decision has been made to purchase a unit.

1.   **How long has the company been in business, and is there a list of referrals the consumer can contact?**

     Check with the Better Business Bureau to determine if any complaints have been received.

2.   **Have the product and the manufacturer been rated by the National Sanitation Foundation (NSF) or other third party organization? Was the product tested for the specific contaminant in question, over the advertised life of the treatment device (with more than 1 gallon of water), under household conditions (tap water, actual flow rates and pressures)?**

     The NSF, whose function is similar to Underwriter's Laboratory, sets performance standards for water treatment devices. Because companies can make unsubstantiated statements regarding product effectiveness, the consumer must evaluate test results of the device to determine if claims are realistic.[4]

     The Water Quality Association (WQA), a self-governing body of manufacturers and distributors, offers voluntary validation programs and advertising guidelines to its members. However, certification, registration or validation may be misleading. For example, manufacturers may be certified by an organization that uses test conditions ideal for contaminant removal, but not representative of all home conditions.[5]

3.   **Does the water quality problem require whole-house treatment or will a single-tap device be adequate?**

     Although less than one percent of tap water is used for drinking and cooking, some contaminants are as hazardous when inhaled or absorbed through the skin as when ingested (e.g. volatile organic contaminants). Treatment of all the water used in the household may be required. Reverse osmosis and distillation units are connected to a single tap; activated carbon devices can be installed on a single tap or where water enters the house. The device selected depends upon the type of contaminant in question.

4. **Will the unit produce enough treated water daily to accommodate household usage? If a filter or membrane is involved, how often will it need to be changed, and how does the consumer know when that change should take place?**

The consumer must be certain that enough treated water will be produced for everyday use. The maximum flow rate should be sufficient for the peak home use rate. Devices such as activated carbon units, reverse osmosis units, and iron filters need routine maintenance; the homeowner should be fully informed of maintenance requirements.

5. **What are the total purchase price and expected maintenance costs of the device? Will the company selling the device also install and service it, and will there be a fee for labor? Can the consumer perform maintenance tasks, or must the water treatment professional be involved? Will the unit substantially increase electrical usage in the home?**

The consumer must watch for hidden costs such as separate installation fees, monthly maintenance fees, or equipment rental fees. Additionally, the disposal of waste materials, such as reject water, spent cartridges from activated carbon units, and used filters, can add to the cost of water treatment and should be figured into the purchase price. Some devices can be installed by the homeowner.

6. **Is there an alarm or indicator light on the device to alert the consumer to a malfunction? Will the manufacturer include in the purchase price a retesting of the water after a month or two?**

Many units have backup systems or shutoff functions to prevent consumption of untreated water. Testing the water a month after the device is installed will assure the homeowner that the unit is accomplishing the intended treatment.

7. **What is the expected lifetime of the product? What is the length of the warranty period, and what does the warranty cover?**

The warranty may cover only certain parts of a device, so the consumer should be aware of the warranty conditions.[4]

PROTECTION FROM DECEPTIVE SALES PRACTICES

If the consumer has had a thorough water quality analysis and the questions above have been answered, then they are well on their way to making an intelligent decision. However, there are a few additional key points to remember.

In-home Testing

Some sellers advertise in the local media, offering a free in-home test of your drinking water if you call. Although in-home testing may be a legitimate sales tool, some promoters use unsophisticated tests to convince you of the need to purchase their product. For example, they may test only for acidity/alkalinity, water hardness, iron, manganese, and color. None of these indicate the presence of harmful contaminants. Others may test only for chlorine, which, although present in your drinking water, may not be at harmful levels. If a salesperson puts a drop of some unknown liquid into a glass of tap water and a little white "stuff" or precipitate appears at the bottom, they may say that this proves the water is contaminated. The precipitate is actually likely to be minerals, many of which are good for you.[6]

"EPA-Approved"

Filters that use silver or chemically active ingredients must be registered with the EPA as required by the Federal Insecticide, Fungicide and Rodenticide Act (FIFRA). The metal inhibits bacteria (see Chapter 3 for details), which makes it a "pesticide". In these cases the unit will be marked with a number prefaced by "EPA Est. No." or "Reg. No.". However, "registration" does not mean approval. No Federal agency approves water treatment units.

Some dealers also may claim that certain government agencies require or recommend widespread use of water filters in homes or restaurants, or that the government approves a particular unit. This is not true. If you see an EPA registration number on a product label, it merely means that the manufacturer has registered its product with the EPA.

EPA registration of a manufacturing establishment (EPA Est. No.) does not mean that the products manufactured/produced in the establishment are registered. It also does not indicate a recommendation or endorsement of any of the establishment's products by EPA. For information regarding registration of water purifiers, contact:

Antimicrobial Program Branch
Registration Division (H7504C)
Office of Pesticide Programs
U.S. Environmental Protection Agency
Washington, DC 20460

"Free Prize!"

Be aware that water filters sometimes are sold as part of prize promotions, which may not be legitimate. Some companies send out postcards saying that you have been selected to receive a prize, and, to receive further details, instruct you to call a telephone number, usually toll-free. If you call, you may discover that you must purchase a water treatment unit to be eligible for a prize, which may be of little or no value. Sometimes sellers will request your credit card number, saying they need to verify your eligibility for a prize or to bill your account. Be cautious about giving your credit card number over the telephone to someone you do not know. Many consumers who have purchased water treatment units from telephone salespersons have found later that the units do not remove contaminants from the water and that they cannot cancel their orders or return the products to obtain refunds.[1]

IMPORTANT POINTS TO REMEMBER

1.   Sophisticated systems can combine two or more technologies. Thus, the short-comings of one technology might be covered by another. However, this increases both the initial cost of purchase and the operating/maintenance costs over the life of the system.

2.   Neglected maintenance is perhaps the biggest problem related to home water treatment units of all types. Therefore, it is very important to become familiar with the maintenance requirements of any treatment unit you own or may buy. Some units require more maintenance than others: follow the manufacturer's recommendations. Additionally, you can ask whether the dealer or manufacturer of your unit/system offers a maintenance contract.

3.   Regardless of the technologies selected, it is important to remember that each water supply is different and each water treatment product is different. Consequently, one cannot assume that a specific product will do an adequate job: for example, some RO units may remove more lead, some may remove less. It is important that units be selected and tested and/or listed by recognized independent organizations to meet specific needs as shown previously. It is not unusual for verbal claims to exaggerate actual capabilities. In the absence of independent testing and listing, one should carefully review the data and claims for the units under consideration and get a written specific performance guarantee from the seller.[1]

WHERE TO GET FURTHER INFORMATION

Several publications over the past few years have summarized POU/POE treatment devices. Some of these include the following:

- Consumer Reports, pp. 27-43, January, 1990.
- Wastewater Journal, pp. 33-37, May, 1988.
- Protect Yourself - A Magazine for Consumers, pp. 29-43 and 60-63, July, 1988.
- Rodale's Practical Homeowner, pp. 28-39, January, 1987.
- East/West, pp. 46-52, August, 1986.
- Rodale's Newsletter, pp. 20-23, May/June, 1985.
- Consumer Reports, pp. 68-73 and 102, February, 1983.
- Changing Times, pp. 44-46, February, 1981.

For background information on Federal regulations of drinking water or additional sources of information, contact:

- Safe Drinking Water Hotline
  1-800/426-4791 (National Toll Free), or
  202/382-5533 (Washington, DC area)

or

- The U.S. Environmental Protection Agency
  Office of Groundwater and Drinking Water
  Washington, DC 20460

Assistance to small communities in providing safe drinking water is fast becoming a major emphasis of EPA. The following (Table 2) are contacts within EPA regions for small communities faced with contaminated drinking water supplies. Figure 2 describes the EPA Regions. Table 3 list the EPA Regional water supply contacts. Table 4 provides the phone numbers of the state radon offices. Two EPA publications that may be helpful are listed below.

- "Is Your Drinking Water Safe?", Office of Water, (WH-550), United States Environmental Protection Agency, EPA 570/9-89-005, June, 1989.

- "Home Water Treatment Units: Filtering Fact From Fiction", Office of Water, (WH-550), United States Environmental Protection Agency, EPA 570/9-90-HHH, September, 1990.

TABLE 2. EPA SMALL COMMUNITY CONTACTS IN EACH USEPA REGION

Region I
Joan Vizziello
John F. Kennedy
   Federal Building
Boston, MA 02203
(617)835-3414

Region II
Berry Shore
26 Federal Plaza
New York, NY 10278
(212)264-7834

Region III
Lawrence A. Teller
841 Chestnut Street
Philadelphia, PA 19107
(215)597-9072

Region IV
Loretta Hanks
345 Courtland St., N.E.
Atlanta, GA 30365
(404)257-3004

Region V
Suzanne Kircos
230 S. Dearborn St.
Chicago, IL 60604
312)353-3209

Region V
Jon Grand
230 S. Dearborn St.
Chicago, IL 60604
(312)353-2073

Region VI
Phil Charles
1445 Ross Ave.,
12th Floor
Dallas, TX 75270
(214)255-2200

Region VII
Ronald R. Ritter
726 Minnesota Avenue
Kansas City, KS 66101
(913)757-2806

Region VIII
Charles Gomez
999   18th   St.,
Suite 500
Denver, CO 80202
(303)330-7575

Region IX
Virginia Donohue
215 Fremont St.
San Francisco, CA
94105
(415)556-7767

Region X
Floyd Winsett
1200 Sixth Avenue
Seattle, WA 98101
(206)399-1466

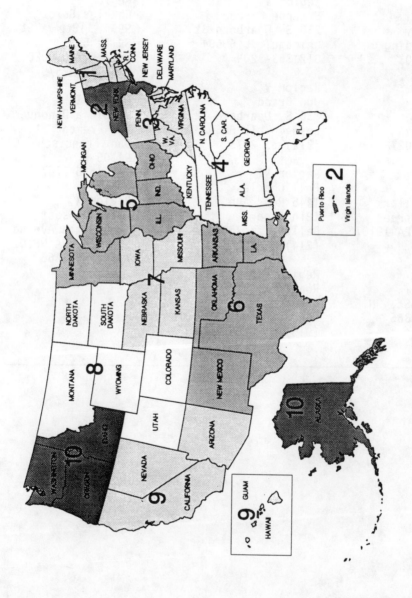

Figure 2. EPA Regional Offices

TABLE 3. EPA REGIONAL WATER SUPPLY CONTACTS

REGION 1
Water Supply Branch
Water Management Division
J. F. Kennedy Federal Building
Room 2203
Boston, MA 02203
617/565-3610

REGION 2
Drinking/Groundwater
  Protection Branch
Water Management Division
Jacob K. Javitz Federal Building
26 Federal Plaza
New York, NY   10278
212/264-1800

REGION 3
Drinking/Groundwater
  Protection Branch
Water Management Division
841 Chestnut Street
Philadelphia, PA   19107
215/597-9873

REGION 4
Water Supply Branch
Water Management Division
345 Courtland Street, N.E.
Atlanta, GA   30365
414/347-4450

REGION 5
Safe Drinking Water Branch
Water Division
230 South Dearborn Street
Chicago, IL   60604
312/353-2151

REGION 6
Water Supply Branch
Water Management Division
First Interstate Bank Tower at
  Fountain Plaza
1445 Ross Avenue
12th Floor, Suite 1200
Dallas, TX   75202-2733
214/655-7150

REGION 7
Drinking Water Branch
Water Management Division
726 Minnesota Avenue
Kansas City, KS   66101
913/551-7032

REGION 8
Drinking Water Branch
Water Management division
999 18th Street, Suite 500
Denver, CO   80202-2405
303/293-1413

REGION 9
Drinking/Groundwater
  Protection Branch
Water Management Division
75 Hawthorne Street
San Francisco, CA   94015
415/744-2125

REGION 10
Drinking Water Branch
Water Division
1200 Sixth Avenue
Seattle, WA   98101
206/399-4092

TABLE 4.   STATE RADON OFFICES

ALABAMA
(205)261-5315
ALASKA
(907)465-3019
ARIZONA
(602)255-4845
ARKANSAS
(501)661-2301
CALIFORNIA
(415)540-2134
COLORADO
(303)331-4812
CONNECTICUT
(203)566-3122
DELAWARE
(800)554-4636
DISTRICT OF COLUMBIA
(202)727-7728
FLORIDA
(800)543-8279
GEORGIA
(404)894-6644
HAWAII
(808)548-4383
IDAHO
(208)334-5933
ILLINOIS
(217)786-6384
INDIANA
(800)272-9723
IOWA
(515)281-7781
KANSAS
(913)296-1560
KENTUCKY
(502)564-3700
LOUISIANA
(504)925-4518
MAINE
(207)289-3826
MARYLAND
(800)872-3666
MASSACHUSETTS
(413)586-7525
OR IN BOSTON
(617)727-6214

MICHIGAN
(517)335-8190
MINNESOTA
(612)623-5341
MISSISSIPPI
(601)354-6657
MISSOURI
(800)669-7236
MONTANA
(406)444-3671
NEBRASKA
(402)471-2168
NEVADA
(702)885-5394
NEW HAMPSHIRE
(603)271-4674
NEW JERSEY
(800)648-0394
NEW MEXICO
(505)-827-2940
NEW YORK
(800)458-1158
NORTH CAROLINA
(919)733-4283
NORTH DAKOTA
(701)224-2348
OHIO
(800)523-4439
OKLAHOMA
(405)271-5221
OREGON
(503)229-5797
PENNSYLVANIA
(800)23-RADON
PUERTO RICO
(809)767-3563
RHODE ISLAND
(401)277-2438
SOUTH CAROLINA
(803)734-4631
SOUTH DAKOTA
(605)773-3153
TENNESSEE
(615)741-4634
TEXAS
(512)835-7000

UTAH
(801)538-6734
VERMONT
(802)828-2886
VIRGINIA
(800)468-0138
VIRGIN ISLAND
(809)774-3320
WASHINGTON
(800)323-9727
WEST VIRGINIA
(304)348-3526
WISCONSIN
(608)273-5180
WYOMING
(307)777-7956

For more information on specific water treatment devices, write or call:

- The National Sanitation Foundation
  3475 Plymouth Road
  P. O. Box 1468
  Ann Arbor, MI   48106
  (313)769-8010

- The Water Quality Association
  4151 Naperville Road
  Lisle, IL   60532
  (312)369-1600

Bottled or bulk water may be a viable alternative to home water treatment if it is needed only for a short period of time. Examples are when a homeowner is pursuing a new source of water or installing a home water treatment system.

Bottled water is regulated as a food by the U.S. Food and Drug Administration (FDA). The FDA concerns itself mostly with sanitation and labeling but is also responsible for ensuring that bottlers comply with Primary and Secondary Drinking Water Standards. Only those bottlers working in more than one state are regulated by the FDA.

Bulk water, which is delivered to the home and stored in large tanks, is available in some communities and may provide a more convenient alternative than bottled water for some people.

For information about bottled water contact:

- Food and Drug Administration
  United States Department of Health and
    Human Services
  5600 Fishers Lane
  Rockville, MD 20857
  (301)443-4166

- International Bottled Water Association
  113 N. Henry Street
  Alexandria, VA 60532
  (703)683-5213

PROBLEM RESOLUTION

To resolve problems concerning a water treatment unit, first try settling any dispute with the company that sold the product. If no satisfaction is obtained in this manner, contact the local consumer protection agency or state Attorney General. Also, one can contact their local Better Business Bureau (BBB). To find the BBB office, check the telephone directory, or write:

•   The Council of Better Business Bureaus
     4200 Wilson Boulevard, Suite 800
     Arlington, VA 22203

To report a complaint about possible sales misrepresentations, write to:

•   The Federal Trade Commission
     Division of Marketing Practices
     Washington, DC 20580

Although the Federal Trade Commission cannot intervene in individual disputes, it is interested in learning about home water treatment sales practices that are believed to be deceptive.

For general water quality information and referrals contact: Your "County Cooperative Extension Service" listed under "County Government" in the white pages of the phone book.

REFERENCES

1.   Federal Trade Commission, "Facts for Consumers: Buying a Home Water Treatment Unit". Office of Consumer/Business Education, Bureau of Consumer Protection, (August, 1989).

2.   Kolega, J. J., "Water Quality Fact Sheet Number 6; Water Testing of Private Wells". The University of Connecticut, Cooperative Extension System, Store, CT, (1989).

3.   "Fit to Drink?", *Consumer Reports*, pp. 27-43, (January, 1990).

4.   Wagenet, L. and Lemley A., "Questions To Ask When Purchasing Water Treatment Equipment". Water Treatment Notes. Cornell Cooperative Extension, New York State College of Human Ecology, Ithaca, NY.

5. Kamrin, M., Christian, B., Bennack, D., and D'Itri, F., "A Guide to Home Water Treatment". Water Quality Series, Extension Bulletin WA21. Cooperative Extension Service, Michigan State University, (January 1990).

6. Lefferts, L. "Water: Treat it Right". *Nutrition Action Health Letter*, 17(9):1,5-7, (November, 1990).